The Implications of Immanence

John D. Caputo, *series editor*

PERSPECTIVES IN
CONTINENTAL
PHILOSOPHY

LEONARD LAWLOR

The Implications of Immanence
Toward a New Concept of Life

FORDHAM UNIVERSITY PRESS
New York ■ 2006

Perspectives in Continental Philosophy Series, No. 56
ISSN 1089-3938

Library of Congress Cataloging-in-Publication Data
  Lawlor, Leonard, 1954–
    The implications of immanence : toward a new concept of life / Leonard
  Lawlor.
      p.   cm. — (Perspectives in continental philosophy ; no. 56)
    Includes bibliographical references and index.
    ISBN-13: 978-0-8232-2653-5 (cloth : alk. paper)
    ISBN-10: 0-8232-2653-0 (cloth : alk. paper)
    ISBN-13: 978-0-8232-2654-2 (pbk. : alk. paper)
    ISBN-10: 0-8232-2654-9 (pbk. : alk. paper)
    1. Life.   2. Immanence (Philosophy).   I. Title.
  BD431.L335   2006
  113′.8—dc22   2006032713

Printed in the United States of America
08 07 06   5 4 3 2 1
First edition

To the Memory of My Parents

Francis J. Lawlor
(1923–2002)

and

Mary A. Lawlor
(1923–1994)

Contents

Acknowledgments

I would like to thank my students and colleagues in the Philosophy Department at the University of Memphis. My thanks, in particular, to all the graduate students who have participated in three of my recent graduate seminars: "Foucault's Early Thought up to *Discipline and Punish*" (Spring 2002), "Merleau-Ponty's Later Thought" (Spring 2004), and "The Problem of Vision in Recent French Thought" (Fall 2004). I must single out Cheri Carr, who edited all of the essays, provided research for many of them, and shared with me many insights concerning the content. The contribution of certain students was particularly important to me: Bryan Bannon, Erinn Gilson, David Gougelet, Marda Kaiser, Heath Massey, John Nale, and David Scott. By accident (it seems) the other continental philosophy faculty at the University of Memphis have also been working on the concept of life; so I must thank Robert Bernasconi and Mary Beth Mader for their kind support of my work and for sharing their own insights with me. It is hard for me to imagine writing these essays in an environment different from the one in the Philosophy Department at Memphis.

Conversations over the past few years with friends (conversations in person, through e-mail, over the telephone, or by means of reading or hearing their work) have been important for the development of the ideas presented in this volume: Keith Ansell Pearson, Gary Aylesworth, Renaud Barbaras, Miguel de Beistegui, Rudolf Bernet, Constantin Boundas, John D. Caputo, Mauro Carbone, Edward S. Casey, Ion Copoeru, Françoise Dastur, Helen Fielding, Linda Fisher, Rodolphe Gasché, Peter Gathje, Galen Johnson, Thomas Khurana, Gary Madison, William McNeil, David Morris, Valentine Moulard, Michael Naas, Thomas Nenon, John Russon, Kas Saghafi, Gary Shapiro, Hugh J. Silverman, Jenny Slatman, Daniel Smith, Ted Toadvine, and Rudi Visker. In particular, I must thank Fred Evans and John Protevi for their kind comments on an earlier draft of this book.

Brien Karas proofread the manuscript and made the index.

I am very grateful to Helen Tartar, who showed interest in this book from the start and who devoted a lot of time and effort to its publication.

As always, I must thank finally Jennifer Wagner-Lawlor for her support.

The essays were written between the spring of 2003 and the autumn of 2005; all have been revised for this book.

"*Verstellung*' ('Misplacement'): Completions of Immanence" first appeared in *The Journal of the British Society for Phenomenology* 36, no. 2 (May 2005): 220–29. I would like to thank Ulrich Haase, the editor of *The Journal of the British Society for Phenomenology*, for allowing me to reprint the essay here. The essay was written for a "book session" on my *Derrida and Husserl* (Bloomington: Indiana University Press, 2002) at the Annual Meeting of the Society for Phenomenology and Existential Philosophy, which took place on November 6, 2003, in Boston, Massachusetts. Commentators on *Derrida and Husserl* were Professors James Mensch of Saint Francis Xavier University and Burt Hopkins of Seattle University. Mensch's essay can be found in *The Journal of the British Society for Phenomenology* 36, no. 2 (May 2005): 208–19; Hopkins's essay appears in *International Journal of Philosophical Studies* 12, no. 2 (2004): 197–218. The version of the essay found here is revised in order to respond to Joshua Kates's critical review of *Derrida and Husserl* in *Husserl Studies* 21 (2005): 55–64. Placing this essay first is intended to indicate that one can find a continuous path from *Derrida and Husserl* to *The Implications of Immanence*.

"With My Hand over My Heart, Looking You Right in the Eyes, I Promise Myself to You . . . : Reflections on Derrida's Interpretation of Husserl" was first published in *Husserl and the Logic of Experience*, ed. Gary Banham (New York: Palgrave MacMillan, 2005), pp. 255–

74. I would like to thank Gary Banham and Palgrave MacMillan for allowing me to reprint the essay here.

"'For the Creation Waits with Eager Longing for the Revelation': From the Deconstruction of Metaphysics to the Deconstruction of Christianity in Derrida" first appeared in a Derrida memorial issue of *Epochē*. I would like to thank Kas Saghafi, Pleshette DeArmitt, and Walter Brogan for allowing me to reprint the essay here.

"Eschatology and Positivism: The Critique of Phenomenology in Derrida and Foucault" first appeared in *Bulletin de la Société Américaine de Philosophie de la Langue Française* 14, no. 1 (Spring 2004): 22–42. I would like to thank Daniel Smith for allowing me to reprint the essay here.

"*Un écart infime* (Part I): Foucault's Critique of the Concept of Lived-Experience (*Vécu*)" first appeared in *Research in Phenomenology* 35 (2005): 11–28. I would like to thank John Sallis for allowing me to reprint the essay here.

A version, with the title "Man and His Doubles: Merleau-Ponty's Mixturism," of "*Un écart infime* (Part II): Merleau-Ponty's 'Mixturism'" first appeared in *Between Description and Interpretation: The Hermeneutic Turn in Phenomenology*, ed. Andre Wiercinski (Toronto: The Hermeneutic Press, 2005), pp. 125–38. I would like to thank Andre Wiercinski for allowing me to reprint the essay here in an expanded version.

"*Noli me tangere*: A Fragment on Vision in Merleau-Ponty" appears here for the first time, but is based on my review essay called "'Noli me tangere': Reflections on Vision Starting from John Russon's *Human Experience*," in *Continental Philosophy Review*.

"*Un écart infime* (Part III): The Blind Spot in Foucault" first appeared in *Philosophy and Social Criticism* (special issue on Foucault) 31, nos. 5–6 (2005): 639–59. I would like to thank Edward McGushin and James Bernauer for allowing me to reprint the essay here.

"'This Is What We Must Not Do': The Question of Death in Merleau-Ponty" appears here for the first time.

"Metaphysics and Powerlessness: An Introduction to the Concept of Life-ism" appears for the first time here. But it is based on a text I wrote for the *Edinburgh University Press Companion to Twentieth-Century Philosophies*, edited by Constantin Boundas.

Abbreviations

Page numbers are first to the original language, then to the English translation. Translations have been modified where necessary to better match aspects of the text under discussion.

CP   Michel Foucault. *Ceci n'est pas une pipe*. Paris: Fata Morgana, 1973. English translation by James Harnes as *This Is Not a Pipe* (Berkeley: University of California Press, 1983).

DH   Leonard Lawlor. *Derrida and Husserl: The Basic Problem of Phenomenology*. Bloomington: Indiana University Press, 2002.

ED   Jacques Derrida. *L'Écriture et la différence*. Paris: Seuil, 1967. English translation by Alan Bass as *Writing and Difference* (Chicago: University of Chicago Press, 1978).

GA9   Martin Heidegger. *Gesamtausgabe*. Vol. 9: *Wegmarken*. Frankfurt am Main: Klostermann, 1976. English translation edited by William McNeill as *Pathmarks* (Cambridge: Cambridge University Press, 1998).

GA29/30   Martin Heidegger. *Gesamtausgabe*. Vol. 29/30: *Die Grundbegriffe der Metaphysik: Welt — Endlichkeit — Einsamkeit* (Frankfurt am Main: Klostermann, 1983). Translated by William McNeill and Nicholas Walker as *The Fundamental Concepts of Metaphysics: World, Finitude, Solitude* (Bloomington: Indiana University Press, 1995).

HL   Maurice Merleau-Ponty. *Husserl at the Limits of Phenomenology*. Trans. Leonard Lawlor with Bettina Bergo. Evanston, Ill.: Northwestern University Press, 2001.

HS1   Michel Foucault. *Histoire de la sexualité, I: La Volonté de savoir*. Paris: Gallimard, 1976. English translation by Robert Hurley as *The History of Sexuality: Volume I: An Introduction* (New York: Vintage, 1990).

Hua III. 1   Edmund Husserl, *Ideen zu einer reinen Phänomenologie und phänomenologischen Philosophie: Erstes Buch*. Ed. Karl Schuhmann. The Hague: Martinus Nijhoff, 1976. English translation by F. Kersten as *Ideas pertaining to a Pure Phenomenology and to a Phenomenological Philosophy*. The Hague: Martinus Nijhoff, 1982. See also Edmund Husserl, *Idées directrices pour une phénoménologie*, trans. Paul Ricœur (Paris: Gallimard, 1950).

Hua IX   Edmund Husserl, *Phänomenologische Psychologie*. The Hague: Martinus Nijhoff, 1962. English translation by Richard E. Palmer in *The Essential Husserl*, ed. Donn Welton (Bloomington: Indiana University Press, 1999).

IP   Maurice Merleau-Ponty. *L'Institution, la passivité, Notes de cours au Collège de France (1954–1955)*. Paris: Belin, 2003.

LOG   Edmund Husserl. *L'Origine de la géométrie*. Trans. and introd. Jacques Derrida. 1962; Paris: Presses Universitaires de France, 1974. English translation by John P. Leavey, Jr., as *Edmund Husserl's Origin of Geometry: An Introduction* (1978; Lincoln: University of Nebraska Press, 1989).

LT   Jacques Derrida. *Le Toucher—Jean-Luc Nancy*. Paris: Galilée, 2000.

MC   Michel Foucault. *Les Mots et les choses*. Paris: Gallimard, 1966. Anonymous English translation as *The Order of Things* (New York: Random House, 1970).

MdA   Jacques Derrida. *Mémoires d'aveugle: L'Autoportrait et autre ruines*. Paris: Editions de la Réunion des musées nationaux, 1990. English translation by Pascale-Anne Brault and Michael Naas as *Memoirs of the Blind: The Self-Portrait and Other Ruins* (Chicago: University of Chicago Press, 1993).

N   Maurice Merleau-Ponty. *La Nature, notes cours du Collège de France*. Ed. Dominique Séglard. Paris: Seuil, 1995. English translation by Robert Vallier as *Nature: Course Notes from the Collège de France* (Evanston, Ill.: Northwestern University Press, 2003).

NC   Michel Foucault. *Naissance de la clinique*. Paris: Presses Universitaires de France, 1963. English translation by A. M. Sheridan Smith as *The Birth of the Clinic: An Archaeology of Medical Perception* (New York: Vintage, 1973).

NdC 1959–61   Maurice Merleau-Ponty. *Notes de cours, 1959–1961*. Paris: Gallimard, 1996.

NW   Martin Heidegger. "Nietzsches Wort 'Gott ist tod.'" In *Holzwege* (Frankfurt am Main: Klostermann, 2003), pp. 209–67. English translation by William Lovitt as "The Word of Nietzsche: 'God Is Dead,'" in *The Question concerning Technology and Other Essays* (New York: Harper, 1977), pp. 53–112.

OE   Maurice Merleau-Ponty. *L'Œil et l'esprit*. Paris: Gallimard, 1964. English translation by Galen John and Michael B. Smith as "Eye and Mind," in *The Merleau-Ponty Aesthetics Reader* (Evanston, Ill.: Northwestern University Press, 1993).

PhP   Maurice Merleau-Ponty. *Phénoménologie de la perception*. Paris: Gallimard, 1945. English translation by Colin Smith, revised by Forrest Williams, as *Phenomenology of Perception* (Atlantic Highlands, N.J.: The Humanities Press, 1981).

QRS   Gilles Deleuze. "A quoi reconnaît-on le structuralisme?" In *L'Île déserte et autres textes*. Paris: Minuit, 2002, pp. 238–69. English translation by Melissa McMahon as "How Do We Recognize Structuralism?" in *Desert Islands and Other Texts* (New York: Semiotext(e), 2004), pp. 170–92.

S   Maurice Merleau-Ponty. *Signes*. Paris: Gallimard, 1960. English translation by Richard C. McCleary as

	Signs (Evanston, Ill.: Northwestern University Press, 1964).
SC	Maurice Merleau-Ponty. *La Structure du comportement.* Paris: Presses Universitaires de France, 1990. English translation by Alden L. Fisher as *The Structure of Behavior* (Pittsburgh: Duquesne University Press, 1983).
SP	Michel Foucault. *Surveiller et punir.* Paris: Gallimard, 1975. English translation by Alan Sheridan as *Discipline and Punish* (New York: Vintage, 1995).
VI	Maurice Merleau-Ponty. *Le Visible et l'invisible.* Paris: Gallimard, 1964. English translation by Alphonso Lingis as *The Visible and the Invisible* (Evanston, Ill.: Northwestern University Press, 1968).
VP	Jacques Derrida. *La Voix et le phénomène.* Paris: Presses Universitaires de France, 1967. English translation by David B. Allison as *Speech and Phenomena* (Evanston, Ill.: Northwestern University Press, 1973).
VS	Michel Foucault, "Vie: Experience et vie." In *Dits et écrits, IV* (Paris: Gallimard, 1994), pp. 763–76. English translation by Robert Hurley as "Life: Experience and Science," in *Essential Works of Michel Foucault: Aesthetics, Method, and Epistemology,* vol. 2, ed. James D. Faubion (New York: The New Press, 1998), pp. 465–78.
WM	Martin Heidegger. "Was ist Metaphysik?" in *Wegmarken* (Frankfurt am Main: Victorio Klostermann, 1967), pp. 103–22. English translation by David Farrell Krell, as "What Is Metaphysics?" in *Pathmarks,* ed. William McNeill (Cambridge: Cambridge University Press, 1998), pp. 82–96.

The Implications of Immanence

Introduction
Signs

Since Marx has become a specter,[1] the signs have become more distinct: we are able to see that we live in an epoch of bio-power.[2] Biopower produces endless contradictions and paradoxes. A deadly epidemic is allowed to develop in one population—the AIDS epidemic in Africa—and at the same time the life of one paralyzed individual is preserved indefinitely—the 2005 case of Terri Schiavo in the United States.[3] Arab "suicide bombers" engage in a kind of auto-affection—auto-affection being the most traditional definition of life itself—that seems to result in no increase of life at all: the suicide bomber affects himself or herself in order to end not only all of his or her own auto-affection but also the lives of many others.[4] Yet in the "East" (to use an easy and questionable designation), the suicide bomber has a different name: he or she is a martyr (a *shahid*).[5] Does this difference in names indicate that the "East" resides in a regime of thought different from that of Western biopower? We can answer this question only with a "perhaps." The unrest in the Darfur region of the Sudan—this clash, of course, is not exactly a clash between the East and the West—shows us something new. Arab warriors there do not rape African tribeswomen only in order to humiliate their enemy; they rape in order to change the blood lines of the tribes. Raped, the African mothers, the "Janjaweed" warriors say, will now bring forth Arab "light" children.[6] Perhaps this new form of violence results from a concern to preserve and increase the power of one population over another. Perhaps even here we have bio-power. How are we to understand these signs (preservation, destruction, and reproduction) if not through a renewal of the concept of life?

Besides these political signs, which call us to conceive life once again, there are philosophical signs, as well. The signs can be found in the final writings of Derrida, Deleuze, and Foucault: Deleuze's "Immanence: a Life,"[7] Foucault's "Life: Experience and Science,"[8] and Derrida's "The Animal That Therefore I Am (More to Follow)."[9] Derrida reminds us that life is always defined by irritability, spontaneity, auto-affection.[10] Deleuze tells us that "a life contains only what is virtual."[11] And, finally, Foucault tells us that there are two approaches to the concept of life: lived-experience and the living (*le vécu* and *le vivant*).[12] But these final texts were written in an attempt to overcome Heidegger's thought of being. We must assert the immense and unquestionable importance of Heidegger. It is precisely Heidegger who called us to overcome of metaphysics. In the Introduction that Heidegger added in 1949 to the 1929 address "What Is Metaphysics?" he says, "From its beginning to its completion, the propositions of metaphysics have been strangely involved in a persistent confusion of beings and being [*in einer durchgängigen Verwechslung von Seiendem und Sein*]."[13] This comment implies that we must dispel the "persistent confusion" between being and beings. In other words, we must aim to differentiate the mixture that metaphysics has made between being and beings. For Heidegger, even Nietzsche has not established the ontological difference; Nietzsche's metaphysics of the will to power is the last version of metaphysics. Heidegger's interpretation of Nietzsche showed us that the will to power aims, through representation, to preserve and enhance power itself. Nietzsche's metaphysics, in Heidegger's interpretation, is the one that we must criticize if we are to exit the regime of bio-power. Heidegger showed us, as well, that we can make this escape only by a form of thought which is nonrepresentational, which is pre- or subrepresentational. Being prior to or below representation, this thought must be one that is outside of representation, that is, outside itself; it must therefore be a thought that is of singularity, that is singular. Yet, thanks to Deleuze, we see distinctly that Heidegger was not able to think singularity: Heidegger's thinking starts from Being and not from the multiplicity of beings.[14] Thanks to Derrida too, we see that Heidegger fails to think singularity: Heidegger's thinking starts from gathering and not from the dissemination of beings.[15] If we are to overcome representation, if we are to overcome metaphysics, then singularity is precisely what calls for thinking. Aristotle said in the *Metaphysics* that there can be no science of the singular; there can be no definition of *ousion ta kath'hekasta* (1039b20). Yet the renewal of the concept of life presented here aims precisely at the formation of that science, or, more precisely, at the formation of that counter-science.[16]

In the formation of what I have come to call "life-ism," an imperative from Merleau-Ponty must guide us. It seems to me that even today, fifty years later, we must obey this imperative if we want to conceive an *archē*, an origin or a principle, such as the principle of life. Here is the imperative in its negative form: the principle must be conceived as neither positive nor negative, as neither infinite nor finite, as neither internal nor external, as neither objective nor subjective; it can be thought through neither idealism nor realism, through neither finalism (or teleology) nor mechanism, through neither determinism nor indeterminism, through neither humanism nor naturalism, through neither metaphysics nor physics. Veering off into one of these extremes is precisely "what we must not do."[17] In short, for Merleau-Ponty, there must be no *separation* between the two poles. But also, there must be no *coincidence*. Neither Platonism (separation) nor Aristotelianism (coincidence) is adequate. The positive formula for Merleau-Ponty's imperative would be the following: instead of a either a separation or a coincidence, there must be "a hiatus," *un écart*, which *mixes* the two together.[18] It seems to me that even today, fifty years later, the mixture that Merleau-Ponty discovered, with its *minuscule hiatus, un écart infime*,[19] is still what requires thinking. This minuscule hiatus, which is death itself, *la place du mort*,[20] defines life-ism. Yet, as we shall see, the only way into the nonplace that defines life-ism lies in the completion of immanence. Indeed, the guiding question throughout the book that follows is: What are the implications of completing immanence?

Verstellung ("Misplacement")
Completions of Immanence

In Derrida, there is a double necessity between an indefinite series
of opposites, such as presence and absence, genesis and structure,
form and content, law and arbitrariness, thought and unthought, empirical and transcendental, origin and retreat, foundation and
founded, and so on. There is a necessity, for example, that genesis
not be separated from structure. It seems to me that we should keep
in mind that, each time we use this double "genesis and structure,"
we are alluding to Jean Hyppolite. In *Logic and Existence*, Hyppolite
equates Hegel's "transformation of metaphysics into logic" with
Nietzsche's statement "God is dead," and this statement, we know,
announces the overcoming of metaphysics as Platonism.[1] Hyppolite
then concludes his discussion of the transformation by saying simply
that "immanence is complete."[2] When I was writing *Derrida and Husserl: The Basic Problem of Phenomenology* (over a ten-year period from
1992 to 2002),[3] I let this Hyppolitian statement guide me. I am still
trying to follow it in everything that I am going to say in this chapter.
Perhaps everything that I have written throughout this book
amounts to an attempt to follow the idea that "immanence is complete." In any case, *Derrida and Husserl* did not concern the relation of
Derrida's thought to that of Hegel. As the title indicates, it concerned
the relation of Derrida's thought to Husserl's phenomenology. Husserlian phenomenology, like that of Hegel (and Hyppolite was aware
of this), is one way of completing immanence, and therefore it over-

comes Platonism: the primary implication of completing immanence. But, perhaps, there are other ways to complete immanence and thus other implications beyond the overcoming of Platonism. Derrida is a phenomenologist, but he is also something else.

In *Derrida and Husserl*, I was able to claim that Derrida is something other than a phenomenologist because I isolated the central structure of Derrida's thought. Indeed, I invoked Derrida's voice, the "unity" of his voice in a kind of univocity. On the basis of a clear understanding of this double necessity of genesis and structure, one might be able to turn it on my own interpretation, arguing that I have made structure and univocity dominate over genesis and equivocity, and that the double has been elided in favor of unity. In fact, one might ask me for an explanation of how I can present Derrida's thought univocally and in a unified and structural way when this very structure, which includes genesis and equivocity, makes such a presentation impossible. One obvious explanation is that I followed certain well-known hermeneutical rules of context reconstruction that I learned from studying Ricœur when I was younger.[4] But a less obvious explanation is that Derrida's thought itself, as far as I can understand it, justifies this emphasis on univocity—it justifies univocity up to a certain point, up to the point of paradox. One of the more remarkable things that I learned in writing this book is that there is a continuity in Derrida's thinking over the last fifty years. Thus, even the final form of his thought—still to be entirely assessed due to his recent passing—revolves around this comment, which comes from *The Problem of Genesis in Husserl's Philosophy* (dating from the academic year 1953–54): "The absolute beginning of philosophy must be essentialist."[5] Derrida's thinking, even his more recent thinking, is based on a kind of essentialist thinking; this is one of the reasons why he uses the French idiom *il faut* as a technical term. In Derrida, *il faut*, "it is necessary," refers to an essential necessity. *Il faut que*, "it is necessary that, it is essentially necessary that," essence and fact are always in principle and in fact inseparable. It seems to me that this sentence—"it is necessary that essence and fact are always in principle and in fact inseparable"—involves a kind of redundancy, or even tautology, between "necessary," "essence," and "in principle," because what "necessary," "essence," and "in principle" refer to is the same. And yet in this sameness there is a *défaut*, a "default" or "defect" that infects the necessity with contingency. The necessity and contingency in *il faut* make it just as paradoxical as another phrase that, more recently, Derrida has coined: "tout autre est tout autre," "every other is every other." The tautological nature of "tout autre est tout autre" and the sameness of "it is necessary that essence and fact are always in principle and in fact inseparable" mean that alterity and necessity, in Derrida, are being thought entirely from themselves, which means that they are based on nothing else, such as an otherworldly idea, substance, or subject. To speak paradoxically, in Derrida, alterity and necessity are being thought on the basis of "the nothing." Therefore, while being fundamental, what is expressed in these two phrases refers us to a nonfoundation, to a *défaut*, to a lack, to an abyss, or to nothing. Derrida is calling us, but in a way that we can understand (which would therefore have to be univocal); he is calling us to think, precisely, the paradox of this tautology, of this *il faut*, of this abyssal foundation. This call to the edge of the abyss cannot be found anywhere in the history of metaphysics—unless one engages in a deconstruction or destruction of that history.[6]

In *Derrida and Husserl*, I said little about deconstructive methodology.[7] In a note I referred to the countless essays and books written over the last twenty-five years or so that describe the so-called "two phases of deconstruction" (DH 2n6).[8] I did, however, claim that Derridean deconstruction is a form of critique (DH 3). If we return to the so-called "two phases," we see that the first phase consists in reversing a hierarchy found in a metaphysical text. So, early in his career Derrida reversed the hierarchy between speech and writing. But to make this reversal come about, Derrida had to show that speech is not autonomous, that it relies in an essential way on writing, and being dependent on writing, speech is always infected with contingency. In other words, "the first phase of deconstruction," as we used to say, engaged in a kind of standard phenomenological critique. Derrida showed that the claims made across the Western tradition, in Plato's *Phaedrus*, for example, that speech precedes and conditions writing are not grounded in evidence, even in intuitional evidence. In this first phase, Derrida could have invoked Husserl's "principle of all principles" against Plato. Yet what about the second phrase? Derrida tells us that the second phase consists in reinscribing (more simply, redefining) the previously suppressed term—in this case, writing—in such a way as to make it designate the origin of the hierarchy. So, based on a first moment of evidential critique, the second in fact concerns genesis, "the basic problem of phenomenology." Yet, this origin—after reinscription, it is called archi-writing—is not transcendent. Deconstruction aims to make us experience it. We are made to experience the essential connection—itself undecidable—between essence and fact, between presence and nonpresence. Insofar as we are led to experience nonpresence, the nonfoundation, we have now moved from an evidential or intra-phenomenological critique to one that is outside of phenomenology ("superphenomenological"). I still think that Derrida's move to the originary experience of nonpresence is new in the history of metaphysics. Has anyone before Derrida really said, "tout autre est tout autre"?

It is Heidegger, however, who first coined such tautological sentences, for instance, "Language is language."[9] In the Preface to *Derrida and Husserl*, I said, "In its widest scope, this book attempts to reconstruct and reflect upon the Derridean transformation of Heideggerian ontology" (DH 2). If we want to understand Derrida, then we cannot forget Heidegger. In the 1949 Introduction to "What Is Metaphysics?" Heidegger says:

> The ecstatic essence of existence is therefore still understood inadequately as long as one thinks of it as merely a "standing out," while interpreting the "out" as meaning "away from" the interior of an immanence of consciousness or spirit. For in this manner, existence would still be represented in terms of "subjectivity" and "substance"; while, in fact, the "out" remains to be thought [*zu denken bleibt*] as the "outside itself" [*Auseinander*] of the opening of being itself.[10]

This *Auseinander* is the completion of immanence. But here, as in Deleuze, immanence does not mean the "immanence of consciousness or spirit." We can speak of immanence here because what Heidegger is speaking about is an interior of the outside. In other words, immanence is the "outside of one another," the outside that is the abyss. And, as Heidegger says here, this "outside" still "remains to be thought." What Heidegger is calling for, and therefore what Derrida is calling for, it seems to me, is a thought from the outside that is about the outside. Thinking comes from the experience of the abyss. This thought of the outside, which emanates from Heidegger, is the widest scope of *Derrida and Husserl*. But, as you know, the phrase "the thought from the outside" alludes to an essay by Foucault, "The Thought from Outside."[11]

At the time I was completing *Derrida and Husserl* in 2000–2001, I was also completing another book, which appeared in 2003: *Thinking through French Philosophy: The Being of the Question*.[12] There I studied the system of thought that arose in France during the 1960s, the system that includes not just Derrida but also Deleuze and Foucault. The studies that this book presents made me realize that there is a point from which these three thinkers develop — *diffract* is the word I used — a point from which these three diffract into independent thinkers. To put it as simply as possible, the "point of diffraction" is the "completion of immanence," and, therefore, the overcoming of metaphysics as Platonism. All three thinkers realized, probably due to Hyppolite, that phenomenology, in both its Hegelian and Husserlian forms, had already attempted such an overcoming. Therefore what unifies these three — Derrida, Deleuze, and Foucault — around 1968 is a reflection on the historical destiny of phenomenology: Does phenomenology really escape metaphysics, understood as Platonism? Already around 1968, one can see differences among the three thinkers. While, for instance, Derrida in *Voice and Phenomenon* — in *Derrida and Husserl* I stress the importance of using the correct translation of *La Voix et le phénomène*[13] — sees phenomenology as the completion and restoration of the original Greek project of metaphysics, Foucault in *Words and Things* (*Les Mots et les choses*) — the great thinker of historical discontinuity — sees phenomenology as a product only of nineteenth-century thinking.[14] Despite this difference, which is significant, both Derrida and Foucault think that phenomenology does not overcome Platonism. To use Heideggerian language, Husserlian phenomenology remains a mere reversal of Platonism and does not twist free of Platonism.[15] While phenomenology is a thought of the same — which is not identity — a thought of tautology, it remains essentially bound to the concept of lived-experience (*Erlebnis*, *vécu*). Indeed, all three — Derrida, Deleuze, and Foucault — present critiques of the central phenomenological concept of lived-experience. In lived-experience, the same is conceived as a *mixture*, as a mixture, for example, of empirical content and foundational form (cf. MC 332 / 321). This mixture means that *sometimes* phenomenology seems to restore Platonism (the foundational forms), *while at other times* it seems to reverse Platonism into its opposite (the empirical content). This mixture of empirical content and foundational form is why chapter 9 of *Words and Things* is called "Man and His Doubles." The philosophical importance of this chapter cannot be overestimated. Yet Foucault gives his most succinct statement of the critique of phenomenology at the end of chapter 7, which is called "The Limits of Representation." Foucault says:

> Undoubtedly, it is not possible to give empirical contents transcendental value, or to move them onto the side of a constituting

subjectivity, without giving rise, at least silently, to an anthropology, that is, to a mode of thought in which the in principle limits of knowledge [*connaissance*] are *at the same time* [*en même temps*] the concrete forms of existence, precisely as they are given in that same empirical knowledge [*savoir*]. (MC 261 / 248, my emphasis)[16]

Even if phenomenology is transcendental, Foucault is saying, it still falls prey to a "silent anthropology." It takes *my* present or *our* present experiences, which are content, as foundational forms. In other words, on the basis of the empirical contents given to *me*, or, better, to *us*, phenomenology tries to determine the form of that empirical content. While trying to keep them separate, phenomenology makes the transcendental and the empirical the same.

Yet in this sameness, for Foucault, there is, as he says, "un écart infime, mais invincible"; the English translation renders this as a "hiatus, minuscule and yet invincible" (MC 351 / 340). Thus, when Foucault infamously calls for the end of man, which he connects with the end of metaphysics (MC 328 / 317), what he is in fact calling for is a *dissociation* of the doubles, such as the empirical and the transcendental or the equivocal and the univocal or genesis and structure, so that a new distribution, *un partage nouveau* (cf. MC 244 / 231), we might say, a new "partitioning," or even a new "spacing," can be established between them. This spacing is the outside, the abyss *(Abgrund)*, or what Heidegger in the 1949 Introduction to "What Is Metaphysics?" calls the *Auseinander*. This *Auseinander* is one way of completing immanence, while phenomenology's way is *Ineinander*. But with this comment I am jumping ahead a bit. I think Deleuze is right in his book on Foucault that this new distribution that Foucault is calling for depends on a new notion of life.[17] In fact, I think that one can overcome metaphysics only through a new notion of life, even more, only through the notion of the living (*le vivant*).

In his most recent work on animality (dating from 1998), Derrida claims that the question of the living and the living animal "will always have been the most important and decisive question" for him, going back as far as his work on Husserl and phenomenology.[18] In *Voice and Phenomenon*, Derrida had indeed spoken of "the ultra-transcendental concept of life" (VP 14 / 15). Derrida calls life ultra-transcendental because it is more fundamental than the parallelism that Husserl establishes between empirical or psychological life and transcendental life. Here is what Derrida says in the Introduction to *Voice and Phenomenon*:

the strange unity of these two parallels, which relates one to the other, does not let itself be distributed [*partager*] by them and dividing itself joins finally the transcendental to its other; this is *life*. We see very quickly that the sole kernel of the concept of psyche is life as self-relation, whether or not it takes the form of consciousness. "Living" [*vivre*] is thus the name of what precedes the reduction and escapes finally from all the distributions [*les partages*] that the reduction makes appear. But this is because it is its own distribution [*partage*] and its own opposition to its other. . . . This concept of life is then grasped in an agency which is no longer that of pre-transcendental naïveté, in the language of life as it is popularly understood or in biological science. [Thus this concept of life is ultra-transcendental.] (VP 14 / 14–15; Derrida's emphasis)

We must notice several things about this long passage. First, the ultra-transcendental concept of life is, according to Derrida, a unity, but a unity that makes its own distribution, *son propre partage*, its own partition. Thus, this unity is strange, and it even seems to be more a duplicity, a double between the empirical and the transcendental. But, second, because the ultra-transcendental concept of life partitions itself, we must think of it as spatial. Life, for Derrida, is, auto-affection, but now we see that in order to think the self-relation, we must think *espacement*, "spacing." We have returned to Foucault's *un écart infime*, and, indeed, at the end of chapter 5 of *Voice and Phenomenon* Derrida describes the difference between retention and primal impression in the living present as *un écart*. The difference in time is a spacing. We can see now also that, like Foucault, Derrida thinks that the problem with the auto-affection of the living voice (as we see in chapter 6 of *Voice and Phenomenon*) is that it makes close, closes, or even encloses this minuscule hiatus. The auto-affection of the living voice encloses the space by means of hearing oneself speak *at the same time* as one is speaking. It encloses the distance into the same by means of the "at the same time." Perhaps this will sound strange, but I think that thinking in terms of time, in terms of a temporal sameness—temporal sameness defines the mixture—is what, for Derrida as well as for Foucault and Deleuze, holds thought captive in metaphysics, and this kind of sameness, which is not a spacing, is why phenomenology remains in metaphysics, is why phenomenology does not really complete immanence. The sameness, not being identity or difference, could also be called ambiguity, and here it is important to recall Derrida's early rejection (in 1972) of polysemy and ambiguity as ways of characterizing undecidability. But the point that I am trying to make is that phenomenology does not think the abyss between the doubles. This abyss between the doubles is why Derrida stresses repeatedly in *Voice and Phenomenon* that, according to Husserl himself, *Fremderfahrung*, the experience of the alien, is always a *Vergegenwärtigung*, a re-presentation. But, for Derrida, the prefix *Ver* implies the *re* of a distance; it implies an "out of place-ness," we might say, even a *Verstellung*, literally, a "re-placing which is a mis-placing," instead of a *Vorstellung*. This mis-placing is life, and insofar as life is always mis-placed, it goes over the limit and includes death—or memory. To think this limit would require, I believe, an investigation of vision, since vision always depends on distance. We might even say that it would require a history of the eye, *une histoire de l'œil*, which would be *mémoires d'aveugle*.[19]

But here one could plausibly ask: Does phenomenology really think in terms of this kind of sameness? To respond to this question, I am going to take up briefly two discussions that one finds in Husserl's published works. The first is the classical definition of *Erlebnis* and therefore of phenomenological immanence. We find this definition in *Ideas I*, section 36, which is called "Intentional Lived-Experiences: Lived-Experiences in General [*Erlebnis überhaupt*]."[20] To distinguish what he is doing from psychology, Husserl says, "Rather [than a discourse of *real* psychological facts; the word *real* is, of course, important] the discourse here and throughout is about purely phenomenological lived-experiences, that is, their essences, and on that basis, what is 'a priori' enclosed in [*in beschlossen*] their essences with unconditional necessity" (Hua III:1, p. 80).[21] As I just said, that Husserl calls psychological facts "real" is important, because all purely phenomenological lived-experiences are, by contrast *reelle*.[22] What Husserl calls intentional lived-experiences, thoughts in the broadest sense, are *reelle*, *and* they contain "the fundamental characteristic of intentionality," the property of being consciousness of something. But, Husserl says, "within the concrete unity of an intentional lived-experience," there are *reelle* moments, which do not have this characteristic of intentionality; these *reelle* moments are the data of sensation. Here, Husserl has discovered something nonintentional and therefore passive at the very heart of lived-experience, something, we might say, that comes from the outside, and yet he has designated these moments as *reelle*, and thereby as "enclosed in" *das Erlebnis überhaupt*. By means of this *überhaupt* and this *in beschlossen*, we can conclude that, in this classical formulation, sameness, which is not identity but a mixture, defines *Erlebnis*. A different completion of immanence, however, would open this enclosure.

The second text at which I would like to look is Husserl's final version of the 1927 *Encyclopedia Britannica* entry for "phenomenology." As is well known, this text introduces phenomenology through phenomenological psychology. Phenomenological psychology, Husserl says here, has the task of investigating the totality of lived-experience. But more importantly, phenomenological psychology, according to Husserl, is an easier way to enter into the transcendental problem that occurred historically with Descartes, that is, that all of reality, and finally the whole world, are for us in existence and in existence in a certain way only as the representational content of our own representations. Thus everything real has to be related back to us. But this "us" cannot be the psyche, according to Husserl, because the psyche is defined by the mundane sense of being as *Vorhandenheit*. To use a mundane being, *Vorhandenheit*, to account for the reality of the world—which is also *Vorhandenheit*—is circular, and this circularity defines psychologism.[23] In contrast to psychologism, phenomenology claims, according to Husserl, that the parallelism between psychological subjectivity and transcendental subjectivity involves a deceptive appearance (*Schein*) of "transcendental duplication." It seems to me that with this duplication we are very close to what Foucault calls "man and his doubles." We are very close because, while Husserl recognizes some sort of difference here, he does not partition the psychological from the transcendental. Instead, he says that transcendental subjectivity is defined by *Vorhandenheit* too, but not in "the same sense [*im selben Sinn*]."[24] Indeed, Husserl thinks that by saying "not in the same sense" he has eliminated the deceptive appearance and makes the parallelism understandable. This is what he says: "the parallelism of the transcendental and the psychological spheres of experience has become comprehensible . . . as a kind of identity of the interpenetration [*Ineinander*] of ontological senses."[25] This "kind of identity" is "ambiguity [*Zweideutigkeit*]." Here Husserl thinks the *Ineinander*, but not the *Auseinander*. Husserl, or phenomenology more generally, does not think the hollowed-out space in the middle of the knot of the double senses of *Vorhandenheit*. He does not think the abyss in between the senses; he thinks the space of sense (or the space of meaning), but not the space that we would have to call nonsense.[26] Husserl's project is *fundamentally* epistemological, but that foundation is precisely the problem; he does not question the ontological mode of consciousness. Without that ontological question, you cannot encounter the hiatus between the doubles. Lacking this question and therefore thinking in the ambiguity, the historical destiny of phenomenology is that it must be overcome. This claim about overcoming phenomenology brings us back to Hyppolite.

In his inaugural address to the Collège de France in 1969, "The Order of Discourse," Foucault said that Hyppolite's *Logic and Existence* established all the problems that are ours.[27] I think that what Foucault said is still true today—but with a change due to the work of Foucault and Derrida (and Deleuze). We can see this change if we go back to Foucault's *Words and Things*. Here, in chapter 9, "Man and His Doubles," Foucault lays out a kind of genealogy of phenomenology. At the beginning of the nineteenth century, he tells us, there was a dissociation of finitude in the double sense, between empirical content and foundational forms of knowledge. This dissociation was Kant's thought. The dissociation, however, led to what Foucault calls a transcendental aesthetics (the empirical content) and a transcendental dialectic (the foundational forms). The transcendental aesthetics became positivism; the transcendental dialectic became eschatology. During the nineteenth century and at the beginning of the twentieth century, this dissociation between positivism and eschatology came to be associated in two ways: in Marxism and in phenomenology. We can see the association in Marxism insofar as Marxism claimed to give the positive truth of man in conditions of labor and *at the same time* promised a revolutionary utopia. We can see the association in phenomenology insofar as phenomenology speaks of the content of *Erlebnis*, which can be positively described as the truth, and *at the same time* of the fulfillment of a meaning-intention, in other words, the promise of fulfilled truth. For Foucault, this association leads to the ambiguity that defines both Marxism and phenomenology. But must we not recognize that with Foucault and Derrida themselves something else has happened? The association that phenomenology and Marxism made has come unraveled. It is almost as if the doubles that came to be the ambiguity of Husserl's thought, positivism and eschatology, have now themselves become dissociated into the thought of Foucault and into the thought of Derrida. On the one hand, we have Derrida's messianism, which leaps back to the eschatology of the nineteenth century. On the other hand, we have Foucault's "fortunate positivism [*un positivisme heureux*],"[28] which obviously leaps back to the positivism of the nineteenth century. It seems to me that this dissociation gives us a problem that is different from theirs: Is it possible to find a new association of eschatology and positivism that is not ambiguous? I think the solution to this problem lies in the direction of multiplicity. But if a new association *is* possible, it would have to be one that continues to think the spacing that partitions; it would have to be a thought of the outside. And in this regard, it would take its inspiration from the simple statement that Hyppolite gave us: "immanence is complete."

With My Hand over My Heart, Looking You Right in the Eyes, I Promise Myself to You . . .

Reflections on Derrida's Interpretation of Husserl

We should never forget that Foucault's *Words and Things* is contemporaneous with Derrida's *Voice and Phenomenon*. We should never overlook the similarity in the two titles, with their little "and." What does this "and" mean? In chapter 9 of *Words and Things*, "Man and His Doubles," Foucault speaks of "a hiatus, minuscule and yet invincible, which resides in the 'and'" of all doubles, such as "the empirical and the transcendental" (MC 351 / 340). Indeed, whereas in the classical epoch time became the foundation of space, in the modern epoch—that is, for Foucault, our present time—space, a "profound spatiality," a distance, has become the foundation of time. All of the thought that evolves from this moment—the moment, we might say, called "1968"[1]—evolves from this distance. Therefore there is, of course, more than one way to conceive this distance.

The most fundamental principle of Derrida's thought is the phenomenological principle of *Fremderfahrung*, from Husserl's Fifth Cartesian Meditation: I can never have a presentation (a *Gegenwärtigung*) of the interior life, the inside, of another; I can only ever have a representation of it (a *Vergegenwärtigung*). Derrida conceives the profound spatiality, the distance, the *écart* through *Vergegenwärtigung*. In *Voice and Phenomenon*, Derrida uses this phenomenological principle of intersubjectivity to contest the implicit "metaphysics of presence" in Husserl's phenomenology. In other words, the necessary possibility of representation in the experience of the alien always

15

contests phenomenology's "principle of all principles," that is, the principle that "every originary donative intuition, everything [in other words] that is given originarily in person [*leibhaftig*], is a legitimating source of knowledge" (*Ideas I*, § 24).[2] For Derrida, there is no pure intuition, not even in my own lived-experience. Even in my solipsistic sphere of ownness, there is only ever a *Vergegenwärtigung*, and therefore some sort of nonpresence and nonbeing. Derrida has generalized *Vergegenwärtigung* to all experience, indeed, to all life. This "deconstruction of phenomenology as the metaphysics of presence," as Derrida himself would say, is still at work in his most recent writings, in, for instance, his 2000 *Le Toucher—Jean-Luc Nancy* (*Touch, to Touch Him—Jean-Luc Nancy*).[3] Within the context of a lengthy discussion of Nancy's "philosophy of touch,"[4] Derrida, for the first time since 1967, revisits Husserl's phenomenology.

Here I intend to examine the new "deconstruction of phenomenology" found in *Le Toucher*, in the chapter called "Tangent II." I hope to show the continuity between this recent text and *Voice and Phenomenon*. As in chapter 6 of *Voice and Phenomenon*, in "Tangent II" of *Le Toucher* there is a critique of the idea that one can have "a pure experience of one's own body" (LT 201). We should note that, throughout *Le Toucher*, Derrida renders German *Leib* as *le corps propre* ("one's own body"), although *Leib* is commonly translated into French as *la chair* ("flesh"). Derrida's translation shows us what is at stake here. Inside of my own, proper body, there is the contamination of the improper, of what is not my own; there is the contamination of others. Or, as the French would say, "au cœur même de ce qui est de mon propre, les intrus me touchent." Thus, more generally, I want to come to understand the philosophy of the heart that animates this entire book (cf. LT 47–48). This "cardiology" is connected to the movement in Derrida's thinking from a thought of the question (as in the question of being) to the promise (as in the promise of justice), from, in other words, ontology (or phenomenological ontology) to eschatology. Indeed, we are going to say that Derrida's kind of critique of phenomenology always results in eschatology.[5]

Presence and Spacing

The new direction appears already in Derrida's 1962 Introduction to his French translation of Husserl's *The Origin of Geometry*.[6] In his Introduction, Derrida stresses that the mathematical object has always been Husserl's "privileged example" of an object (LOG 6/27).

The mathematical object (and thus the geometrical object; cf. LOG 79n2/83n87) holds a privileged position for Husserl because it is what it appears to be; it is ideal (LOG 6/27); it is pure truth (cf. LOG 66–68/74–75). As Derrida says, the mathematical object "is thoroughly transparent and exhausted by its phenomenality" (LOG 6/27). The mathematical object is, therefore, always already reduced to being a phenomenon. Being a phenomenon, it is permanently available to a "pure look" (LOG 72/78); it is a theorem in the literal sense of something "looked at" (cf. LOG 78/83 and VP 80/72). While being a pure object, the mathematical object is being to a pure consciousness (LOG 6/27). This definition of a mathematical object as being already reduced to its phenomenality, as an ideality related to a pure consciousness or look (*le regard*) means that the being of the mathematical object is a thought-being—or, as Husserl would say, a "noema" (something thought).

In his early but important "'Genesis and Structure' and Phenomenology,"[7] Derrida recognizes that, in *Ideas I*, a series of ontological differences—the difference, in particular, between, real, *reell*, and *ir-reell*—defines the concept of the noema (ED 242–44/162–64). As is well known, consciousness, for Husserl, includes the noema within itself because the noema is not *real* in the sense of being a factual thing. The noema is not a worldly fact but an ideal sense intended by consciousness. Nevertheless, even though consciousness includes the noema, it is not a *reell* property of consciousness. As Derrida says, the noema "is neither of the world nor of consciousness, but it is the world or something of the world *for* consciousness" (ED 242/163). The noema does not *really*—in either sense of the word *real*—belong either to the world or to consciousness, even though it *participates* in both;[8] it is therefore "the root and very possibility of objectivity and sense" (ED 243/163). That the noema does not originate in any region implies, as Derrida says, "an anarchy of the noema" (ED 243/163). Here, Derrida intends "anarchy" not only in the literal sense of not having a principle (*archē*)—an-archy—but also in the more normal sense of a disorderly movement. Because the noema is anarchical—both as having no principle from which it originates and as a movement across many regions without belonging to any of them—it implies a profound spatiality.

We can see that presence involves a profound spatiality if we turn now to the definition of presence that Derrida provides in *Voice and Phenomenon*, chapter 6. If *Voice and Phenomenon* concerns the "ultra-transcendental concept of life," as its Introduction states (VP 14/

15), and if life, for Derrida, is defined by auto-affection, then chapter 6 ("The Voice That Keeps Silent") is the very heart of *Voice and Phenomenon*. Here is the definition of presence: "presence [is] *simultaneously . . . the being-before of the object*, available for a look, and . . . *proximity to self in interiority*. The 'pre' of the *present ob*ject now-before is an *against* [contre] (*Gegen*wart, *Gegen*stand) simultaneously in the sense of the *wholly against* [tout-contre] of proximity and in the sense of the *encounter* [l'encontre] of the op-posed" (VP 83–84 / 75; Derrida's emphasis). Presence, for Derrida, can become a problem because it must be *at once* close by and proximate, and *at once* away and distant. In other words, it must be self-presence and presence, the object as repeatable to infinity and the presence of the constituting acts to themselves. Indeed, the constitution of an ideal object requires a medium that can preserve self-presence and constitute presence. As is well known, this medium (which is the medium of expression) is the voice, the voice that keeps quiet, as in interior monologue, where one hears oneself speak.

Hearing oneself speak is, according to Derrida, an "absolutely unique type of auto-affection" (VP 88 / 78). The phenomenological essence of this self-relation seems to consist in three moments. First, whether I actually use my vocal cords or not, there is forming either in my head or in my mouth the forms of sounds, phonemes; the phonemes are produced in the world (VP 89 / 79). So the voice is nothing outside of the world. Nevertheless, as Derrida stresses, this mundaneity of the sound does not mean that an objective, mundane science can teach us anything about the voice (VP 89 / 79). A mundane science cannot teach us anything about the voice because, second, hearing oneself speak is temporal, that is, the sound is iterated across moments. This temporal iteration is why sound is the most ideal of all signs (VP 86 / 77). Third, in hearing oneself speak, one still exteriorizes one's thoughts or "meaning-intention" or acts of repetition in the phonemes. This exteriorization — ex-pression — seems to imply that we have now moved from time to space. But since the sound is heard by the subject during the time he is speaking, the voice is in absolute proximity to its speaker, "within the absolute proximity of its present" (VP 85 / 76), "absolutely close to me" (VP 87 / 77). The subject lets himself be affected by the phoneme (that is, he hears his own sounds, his own voice) without any detour through exteriority or through the world, or, as Derrida says, without any detour through "the non-proper in general" (VP 88 / 78). Hearing oneself speak is "lived [*vécue*] as absolutely pure auto-affection" (VP 89 / 79).

In this auto-affection, one stays within what Husserl in the Fifth Cartesian Meditation called "the sphere of ownness." What makes it be a pure auto-affection, according to Derrida, is that it is "a self-proximity which would be nothing other than the absolute reduction of space in general" (VP 89 / 79). This absolute reduction of space in general is why hearing oneself speak, the unity of sound and voice (or meaning-intention), is so appropriate for universality (VP 89 / 79). Requiring the intervention of no surface in the world, the voice is a "signifying substance that is absolutely available" (VP 89 /79). Its transmission or iteration encounters no obstacles or limits. The signified, or what I want to say, is so close to the signifier that the signifier is "diaphanous" (VP 90 / 80). Therefore, *"the phoneme gives itself as the mastered ideality of the phenomenon"* (VP 87 / 78; Derrida's emphasis). As Derrida says: "the unity of the sound (which is in the world) and the *phonē* (in the phenomenological sense), which allows the latter to be produced in the world as pure auto-affection, is the unique agency which escapes from the distinction between intra-mundaneity and transcendentality. And by the same token, this unity is what makes the distinction possible" (VP 89 / 79). The voice, therefore, for Derrida is ultra-transcendental. It is the very element and means of consciousness (VP 89 / 79–80).

For Derrida, this voice functions even at the most fundamental, pre-expressive level, at the very root of transcendental *Erlebnis*, in absolute silence. The voice functions here because its primary determination is temporality (VP 93 / 83). According to Derrida, the living present in Husserl, originary temporalization, is "pure auto-affection" (VP 93 / 83). It is pure auto-affection because, as Derrida says, for Husserl temporality is never the predicate of a real being, which means that the intuition of time is a receiving that receives nothing and a production that produces nothing, nothing in the sense of a real being. Therefore, temporalization is a "pure movement" producing itself: spontaneous self-generation (VP 93 / 84). Nevertheless, this transcendental auto-affection supposes that "a pure difference . . . divides self-presence" (VP 92 / 82), the pure difference between receiving and creating (VP 92 / 82). This pure difference originarily introduces into self-presence, according to Derrida, impurity: space, the outside, the world, the body, and so on (VP 95 / 85, 92 / 82). As the hyphen indicates, there could be no self-generation without the possibility of a difference, which divides the self, the *auto* (VP 92 / 82): "It produces the same as the self-relation within the difference with oneself, the same as the non-identical" (VP 92 / 82). This nonidentical sameness is Derrida's famous concept of différance; "The living present springs forth on the basis of its non-self-identity" (VP 95/85).[9] When I *hear* myself speak, the hearing is a repetition of the speaking that has already disappeared; re-presentation (*Vergegenwärtigung*) has intervened, and that intervention means, in a word, space. As Derrida says, "the 'outside' insinuates itself in the movement by which the inside of non-space, what has the name of 'time,' appears to itself, constitutes itself, 'presents' itself" (VP 96/86). Within time, there is a fundamental "spacing [*espacement*]" (VP 96/86). Spacing implies what Derrida calls "archi-writing" and vision: "when I *see* myself writing and when I signify by gestures, the proximity of hearing myself speak is broken" (VP 90/80; my emphasis). The visual surface also breaks the immediate contact of touching oneself.

In 1967, the year of *Voice and Phenomenon*'s publication, the philosophical importance of the touching-touched relation had not yet been widely recognized, given that Merleau-Ponty's posthumous *The Visible and the Invisible* appeared only in 1964. Likewise, the philosophical importance of Husserl's *Ideas II* had not yet been widely recognized. In chapter 6 of *Voice and Phenomenon*, Derrida mentions the touching-touched relation only in passing (VP 88/79). This is what he says:

> When I see myself, either because I gaze upon a limited region of my body or because it is reflected in a mirror, what is outside the sphere of "my own" has already entered the field of this auto-affection, with the result that it is no longer pure. In the experience of touching and being touched, the same thing happens. In both cases, the surface of my body, as something external, must begin by being exposed in the world. (VP 88/79)

This passage shows the great continuity of Derrida's thought from this moment up to the present. Here, Derrida makes vision and touch comparable. Thus, what he says in 2000, in *Le Toucher*, could be applied to *Voice and Phenomenon* in 1967: "Do we dare say, figuratively, that [the difference between vision and touch] has to do [*à voir*] with everything that concerns us here, or that [this difference] hits at [*touche*] everything that is important to us in this book?" (LT 194). As Derrida is aware, the French idioms—*toucher* and *à voir*—reinforce the similarity between vision and touch. But the Western philosophical tradition has not stressed the similarity between these two senses. Instead, touch, and especially touching with the hand and fingers, is supposed to be superior to vision because touch is supposed to be immediate contact without apparent distance. As we can see in the passage from *Voice and Phenomenon* above, Derrida, in contrast to the tradition, has allied touch with vision precisely because of the *mediate* character of both, because of the surface exposed to the world. This surface exposed to the world makes visual or tactile auto-affection impure. In *Le Toucher*, Derrida is doing the same thing; he is trying to show the impurity of tactile auto-affection, that is, that this auto-affection is also hetero-affection. Thus, again, what Derrida says in 2000, in *Le Toucher*, could be applied to *Voice and Phenomenon* in 1967: "It would no longer be a question of granting or restoring a privilege or priority to some one or other sense, vision, touch, hearing, taste, smell. 'There is no touch as such [*"le" toucher*],' . . . and . . . there is no vision as such [*"la" vue*]. Rather, what would be at issue is to reorganize this whole field of sense and of the senses differently. [This reorganization] would no longer depend on a particular sense called touch" (LT 206). Let us see what it would depend on.

Places: The Deconstruction of Husserl's Phenomenology in *Le Toucher*

In the "Avant-Propos" to *Le Toucher*, Derrida tells us that the guiding thread of this book is *le toucher*, in the double sense of a noun, the sense of touch, and a verb, in which *le* would become an object pronoun, "to touch him" or even "to touch at him" (*toucher à lui*; LT 9). But Derrida also says that he has wanted to write (dedicated to Jean-Luc Nancy) a "history" of *le toucher* starting from Aristotle; it could have been called "peri Peri Psykhès": around Aristotle's treatise that we commonly call in English *De Anima*, or at the periphery of or about *De Anima*. The ambiguity that Derrida is suggesting here is this: around the "around" of the psyche or the soul, as if the soul were spatial and not temporal. The privilege of the sense of touch throughout the history of Western thought derives from the *immediacy* or *simultaneity* of contact. What Derrida is trying to show in these histories of *le toucher*, however, is the distance or mediation interrupting the contact.[10] One of Derrida's "histories" is Husserl's famous descriptions of touch in *Ideas II*.[11] Despite this fame, Derrida wonders "how can we ignore the fact that these same texts [in *Ideas II*] are still waiting, with an infinite patience, to begin to be read and re-read again?" (LT 184n1). In "Tangent II," Derrida attempts one of these re-readings of *Ideas II*, in particular, of paragraphs 37 and 45 in Section 2, "The Constitution of Animal Nature." The re-reading takes place in three steps.[12]

First, Derrida identifies in Husserl the signs of the "excellence" of touch among the senses and of the hand among the parts or organs of "one's own body" (*Leib*).[13] In section 37 ("Differences between the Visual and the Tactual Realm"), Derrida focuses on the opening passages. Here, Husserl starts out by speaking of "the external object" that can be touched by the other external object, one's own body, and ends by speaking of the double sensation involved in touching (LT 187).[14] According to Derrida, Husserl's granting of a primacy to the thing as external object (LT 186) *and* the fact that the hand and the fingers can touch one another reflexively explains his privileging of "digital" or "manual" touching. Touching is at once both objective and subjective. Thus, "the primacy of sub-objectivity" explains "the phenomenological nobility" of digital touching (LT 188).

Derrida punctuates this first step of his re-reading with two "let us never forget's" ("n'oublions jamais") and two "even if it is not necessarily justified's" ("même s'il ne se justifie pas nécessairement"). With the first set of punctuations (the two "let us never forget's"), Derrida is trying to make certain that we do not forget what Derrida himself calls his "old concerns [*préoccupations anciennes*]" (LT 206). Therefore, "let us never forget" that, when Husserl in paragraph 37 speaks of "fingers touching fingers," "these fingers are fingers that touch and that do not show, indicate, signal, or signify. The deictic function of the index finger seems here to be reduced to the differentiated potentiality of a contact: I would point the finger at what, at the end of my approaching movement, one could hope to reach and, so to speak, touch" (LT 187). Let us never forget that *Voice and Phenomenon* deconstructs phenomenology by showing that pure expression or pure perception or pure presence is always contaminated, that is, taken or even touched (*pris*) (VP 21 / 20) by the sign, by indication, by the indexical, by "pointing the finger [*montrer du doigt*]" (VP 24 / 23). Pointing the finger includes necessarily the possibility of a visual surface, not immediate contact. The second punctuation comes immediately after: "Let us also not forget that 'sensing oneself touch*ing with the* [du] finger' is immediately a 'sensing oneself touch*ed through the* [du] finger,' even when my finger does not touch another of my fingers, when it touches anything external to my body and my finger, my finger senses that it is touched by the thing that it is touching" (LT 187). In other words, the touching-touched relation is general to touching; it occurs wherever touching occurs, even when it is not one human hand—"my hand," as Husserl says in paragraph 38—touching another hand. This generalization of the relation means that the sense of touch, no matter what, involves passivity: *le toucher*, in its ambiguity between verb and noun. If *le toucher* is a verb, then the *le* becomes either a direct or an indirect object pronoun; *le toucher* then means that someone is touching *at* me, as in pointing the finger at me. Again, let us never forget that *Voice and Phenomenon* deconstructs phenomenology by showing that pure expression is always "touched [*pris*]" by the index finger. The ambiguity of *le toucher* has one further consequence. Derrida is reminding us that the privilege that Husserl gives to the hand, to one hand touching another, does not seem necessarily justified. And this lack of necessary justification brings us to the other set of punctuations that Derrida makes in this first step.

Husserl privileges the hand touching the hand, of course, because the hand can touch another part of one's own body and be touched by it: the double sensation. Derrida acknowledges that one cannot say there are double sensations at every part of the external body. "But why the hand and the finger alone?" Here Derrida presents a list of the most obvious external body parts in which the double sensation is possible: the feet and toes, the lips, the tongue and the palate, the eyelids, the anal and genital orifices. By choosing the hand and the fingers, Husserl has changed the stakes of the description. Instead of the immediacy of tactile perception, the mediation of language could have been at stake. As Derrida points out, the tongue, the palate, and the lips touching one another are all required for speaking (LT 188). Thus, the privilege of the hand is explained (by the double sensation), "even if it is not necessarily justified" (LT 188). But also, the choice of the human hand "is not always justified" (LT 191), since many animals have organs or members that resemble the human hand (LT 193). Nevertheless, Husserl's choice of man, according to Derrida, is explained in three ways. First, it is possible to select man and constitute a phenomenological anthropology. Second, the intuitionism of phenomenology requires that we begin with "us," with what is closest and most one's own, without any indirect representation, even if this starting point in us falls short of the most radical phenomenological reduction. And third, Husserl's choice of man is explained by "an evaluation, a teleological philosophy of life" (LT 191). Derrida provides a "proof" of this teleology by going to the beginning of paragraph 43, where Husserl speaks of social relations being instituted among men, and then he adds, in parentheses, that social relations are made "already among animals to a lower degree" (LT 192). "Lower degrees" (*niederster Stufe*) here indicates a hierarchy that, Derrida says, is "classical: matter, life, mind" (LT 192). Thus once again Husserl's privilege of the hand is difficult to analyze and to justify. But Husserl's selection of the hand seems to be commanded by "two heterogeneous, if not contradictory imperatives" (LT 193), and these imperatives were already anticipated by the sub-objectivity mentioned above. The contradiction is that Husserl must speak about one's own body by means of perception of external objects (which implies, as we shall see, that there will also be internal objects within one's own body), but he also must describe the subjective experience of reflexivity (LT 193). According to Derrida, it is on the basis of this double subjective-objective imperative that Husserl differentiates between touch and vision.

This differentiation brings us to the second step in Derrida's re-reading of *Ideas II*, §37 and §45. In §37, Husserl says, "we immediately notice the difference" between the eye and the hand, between vision and touch.[15] As Derrida stresses, for Husserl, the eye differs from the hand because the eye lacks something.[16] The French translation of *Ideas II*, which Derrida is using, translates *es fehlt* by *ce qui manque*, and Derrida calls this "lack" (*manque*) "un défaut" (LT 195).[17] There is a "defect" in vision that does not seem to affect touch. This defect is, as Husserl says, that "the eye does not appear visually." In other words, the eye does not see itself, at least not directly or immediately, whereas the hand seems to touch itself, directly or immediately. In a note, Husserl considers the possibility of a mirror, but stresses that, when I look in the mirror, I do not see *my* eye; rather I see my eye as a thing as I see the eye of another. For Derrida, Husserl's description of the specular experience implies that the indirection of seeing one's own eye includes necessarily mediation, analogical *Vergegenwärtigung* (*Einfühlung*), and technology (LT 195–96). By contrast, the immediacy of touch, for Husserl, has to do with the localization of sensations right on the hand. As Derrida says, paraphrasing Husserl, "The eye does not see, and the sensations of colors are not localized right on [*à même*] the seeing eye, or right on [*à même*] an eye appearing 'visually,' as would be the case of the touched object perceived right on the touching hand"(LT 196).[18] So, the difference between the eye and the hand concerns the immediacy of localization. Concerning localization, Husserl in §37 makes a division between two topological experiences. On the one hand, there are the sensible impressions located on the surface of the hand and inside of it; on the other, there is experience of the extension of the material determinations of the thing (its roughness, for example). According to paragraph 37, then, there is a phenomenological surface and an interiority to the hand, and a surface and exteriority to the real thing. The sensible impression on the phenomenological surface of the hand and inside it is, as Husserl says, "the hand itself, which for us is more than a material thing." But this division in superficial (in the literal sense) experiences leads to a further consequence: "the way that [the hand] belongs to me entails this consequence that I, 'the subject of the body,' says that what belongs to the material thing is its, not mine."[19] In other words, the way the hand belongs to me allows me to differentiate between myself and another.

Derrida sees in this "place" (*lieu*)—in the surface of the hand—one of the zones in which phenomenology encounters the strongest resistance to the authority of its intuitionistic "principle of all principles" (LT 198).[20] Now, Derrida stresses that the principal criterion of the difference within the surfaces is that "the real qualities are constituted by means of a 'sensible schema' and a 'multiplicity of adumbrations,' while the tactile, sensible impressions imply neither adumbrations nor schematization" (LT 198). Derrida calls this distinction a "properly phenomenological necessity" (LT 199). We can see in this distinction, of course, both the reduction and the distinction between real and *irreell*, "the double motif that is classic in Husserl" (LT 257).[21] But the *irreell* nature of the sensible impression—its sense—implies that it does not belong to me, as do the *reell* components. This nonadherence of the sense of the contact implies that it is heterogeneous and external to the sense impression. Or, even if the sense impression is *reell* as the hyletic content of the tactile lived-experience, it too, as a passive givenness, as not being intentional or morphic, comes from the outside (ED 243 / 163). As Derrida says, "a certain exteriority, an exteriority that is heterogeneous to the sensible impression and that is even real (and even, as Husserl recalls, a 'real optical property of the hand'[22]) *participates*; an exteriority perceived *as* real *must* even participate in the experience of touching-touched, and of the 'double apprehension,' be it by virtue of the hyle-morphe relation or by virtue of the noetic-noematic relation" (LT 200). This exteriority is necessary in order for the double apprehension to be double. Without this outside, the double apprehension would be an apprehension of only the same. "This detour through the foreign outside" is also what allows me, the one undergoing this singular experience of the double apprehension, to distinguish between me and not-me, the ego and the not-ego. We neither must nor can have a coincidence in the contact if the apprehension is to be double. There must or can be a difference, the heterogeneity of a spacing between the touching and the touched. In other words, it must be possible — Derrida is speaking of a necessary possibility, which does not in fact have to happen—for the hand to be seen, and thus the double apprehension "cannot be reduced to a pure experience of what is purely one's own body" (LT 201). This impossibility of having a pure experience of what is purely one's own body takes us to the heart, the third step, of Derrida's re-reading.

In this third step, Derrida moves into the next chapter of *Ideas II*, section 2, in particular into paragraph 45, "Animalia as Primally Present Corporeal Bodies with Appresented Interiority." Here Husserl speaks of *Herzgefühl*, "ich empfinde mein Herz," which is rendered into French as "je sens mon cœur" and into English as "I feel my heart." But like the German verb *empfinden*, in this expression the French verb that translates it, *sentir*, indicates that what is at stake here is sense and the senses. As we can see from the title, this paragraph concerns intersubjectivity and therefore *Vergegenwärtigung*. Derrida tells us that what Husserl says about the heart here is important to him for two reasons. On the one hand, there is an architectonic reason: here Husserl is trying to describe how the "human-ego" is constituted through *Einfühlung*. To do this, Husserl returns to a starting point in solipsistic experience in which there seems to be no *Einfühlung*; *Einfühlung*, for Husserl, comes *later* than the solipsistic experience. This "later" is why Derrida speaks of an architectonic concern; his question is: Where does *Einfühlung* begin? (LT 202). The solipsistic experience gives Derrida a second reason for making what Husserl says about the heart important. Husserl turns to this "I feel my heart" because it is a better example of a solipsistic experience than the touching-touched experience, which is a double apprehension. Derrida stresses that Husserl cannot even speak of the feeling of the heart, since to speak of it already implies intersubjectivity, the address to someone, even if this someone is me (LT 202). The feeling of the heart, of this chamber, is silent. Thus, when Husserl attempts to connect the feeling of the heart back to touch (which is not *solus*, alone), what Derrida sees here is "a certain difficulty" (*un certain embarras*) (LT 202).

Here, in paragraph 45, Derrida again (as in his re-reading of §37) follows Husserl's description very closely.[23] Husserl is describing the different fields or places on the solipsistic subject's body. Below the surface of the body, there is, as Husserl says, "the localization of somatic interiority mediated by localization of the field of touch." Derrida emphasizes the word *mediated* (*vermittelt*), because it is clear that the feeling of my interiority is not immediate. Obviously, I cannot immediately reach inside myself and take hold of my heart. Thus Husserl speaks of touch here to reintroduce immediacy: I cannot immediately or directly sense my heart unless, as Husserl points out, I press the surface of my own body in the region of the heart. In other words, as Derrida says, "With the hand I touch the inside of my body across (*à travers*[24]) a surface" (LT 203). Then we can really speak of touching, *but also* of visibility, since we have a surface. For Derrida, as soon as we have the visible surface, we have mediation and distance: the localization of somatic interiority is indeed mediated. And therefore when I press my hand over my heart, it is, as Husserl says, the same as "with other bodies: I feel through [*durch, à travers*] to their insides." For Derrida, this mediation at work already in the solipsistic experience implies that the possibility of intersubjectivity must be there always already, as if my heart were always already the heart of another. It is not possible to say, as Husserl does in this description, that *Vergegenwärtigung* happens *dann, ensuite*, "then." Invoking the phenomenological insight of the Fifth Cartesian Meditation, we could say that, just as I can only ever have a *Vergegenwärtigung* of the interior life of another, I can only ever have a *Vergegenwärtigung* of the interior life of my own body. I could reach out and feel your heart beating, I could even (given surgery and if I were a "doubting Thomas") reach out and put my hand right on your beating heart, but so long as there is a necessary possibility of the visual surface, there is mediation, which distances and therefore interrupts the contact. We must never forget that the heartbeat is syncopated, like the blinking of the eye. It is like a voice that keeps silent.

Across Positivism and Eschatology

Because of the example of the heart, we can see that this critique (found both here in "Tangent II" of *Le Toucher* and in chapter 6 of *Voice and Phenomenon*) concerns the notion of life. It is a critique of solipsistic experience, or, more precisely, lived-experience. What kind of critique is this? It depends entirely on a necessary possibility: wherever there is sensing, it must be possible for there to be a surface, and wherever there is a surface, it must be possible for there to be space. This necessary possibility implies that solipsistic experience, being alone and therefore close to oneself and unified with oneself, is always already virtually double, distant from oneself, and divided. *But*—this is an important "but," as we shall see in a moment—what divides the *ipse*, spacing it and making it double *in Derrida* is mediation, *Vergegenwärtigung*. In Derrida, mediation contaminates the immediate, but contamination is still mediation. Thus, contamination promises unity even though it cannot, by necessity, ever keep this promise. The other is always already close by and coming without ever arriving. Without ever being able to arrive, the one who is going to keep the promise is to come in person (*Leiblich*). Therefore, we must characterize Derrida's critique of phenomenology (as he himself has done) as an *eschatological* critique. It is a critique based in a promised unity that demands to be done over again and again.[25]

The "discontinuity" in Derrida's thought consists in a shift in his conception of fundamental experience—that is, in his conception of what he called in *Voice and Phenomenon* "the ultra-transcendental concept of life." This is a shift from the model of interrogation to the model of a promise. We can see this shift in *Le Toucher*. The book opens with a question: "One day, yes, one day, once upon a time, extraordinary, addressed one time extraordinarily, addressed with as much violence as tact [*de doigté*], some such question took hold of me [*me prit*], as if it came to me" (LT 11). The question was: "When our eyes meet, is it daytime or nighttime?"[26] When we say, "our eyes meet," we know that it means the beginning of a bond, which could lead to love. Thus, when our eyes meet, this experience "touches" the heart. If *Le Toucher* is indicative of Derrida's recent thought, it seems that he conceives life or the psyche (*De Anima*) on the basis of the heart (LT 48, LT 201–2, LT 319); he says, "nothing appears, at the least, more auto-affective than the heart," "the origin of life" (LT 301).

To conceive life, Derrida exploits the heart's many nuances. The heart is a place, a place that mediates the circulation of blood and keeps the body alive (LT 233). But we also say that we learn something by heart, which means that the heart is the place of memory (LT 47, LT 325). Here, Derrida resurrects all the associations he brought forward with *écriture*. As the place of memory, the heart is a memory device, a machine, and thus technology. As the place of memory, the heart is a crypt, a place of death and mourning. And as the place of death—the heart beats, after all, and there is an interruption in the flow of blood, like death (LT 257)—the heart can be faulty. The heart can betray you; it can *perjure* itself (*se parjurer*). Then you require a heart transplant, the heart of another, an intruder right inside of your own flesh (LT 319). Thus, when we say "I think with my heart," it is possible that I think with the heart of another. For Derrida, we can even say that the heart is thought itself, the commandment of promising (LT 47, LT 325). When I say that "I think with my heart," I experience this thinking as an imperative that does not come from reason but from elsewhere. As in the Sacred Heart of Jesus, overflowing with compassion (*misericordia*; LT 118, LT 281, LT 286), the heart promises to give out blood without the expectation of return (LT 318). This generosity can be seen in the carrying that is pregnancy: there are two hearts (LT 135n1) and the promise of someone coming. Indeed, when our eyes meet, I can see right into your heart and touch it. With all my heart, I promise I will be faithful (LT 301). I will come; I swear (*jure*). *With my hand over my heart, looking you right in the eyes, I promise myself to you.*

3

"For the Creation Waits with Eager Longing for the Revelation"
From the Deconstruction of Metaphysics to the Deconstruction of Christianity in Derrida

Perhaps Derrida's most enduring contribution to philosophy, to thinking in general, is the idea of a "deconstruction of the metaphysics of presence." Unlike that of Heidegger—with which it nevertheless has so much in common—Derrida's deconstruction does not aim at retrieving Being, *Wesen*, or *Anwesen*. I would say that it aims at retrieving the soul, the *psyche*, *anima*. Life is what is otherwise than being.[1] We see this aim at life as early as 1967 in the Introduction to *Voice and Phenomenon*, where Derrida says that "the ultra-transcendental concept of life" is the source of "all the distributions" made in phenomenology (VP 14/00). We also see this aim as recently as 1998, in a text based on his lectures on animality; here Derrida tells us that the question of the living and the living animal "will always have been the most important and decisive question" for him.[2] The Western tradition has always defined life, insofar as it is opposed to the inorganic, as auto-affection.[3] Life is auto-affection. In *De Anima*, or *Peri Psyches*, *On the Soul*, Aristotle already implies this definition through the essential role he gives to touch and movement in the sense of nutrition (413a–413b11).[4] And Derrida never tires of reminding us that the French word *psyché* refers to a large mirror.[5] As we just saw in Husserl's *Ideas II*, a mirror, of course, is that in which one can see oneself seeing. It seems to me that perhaps the principal idea guiding Derrida's thought is that, wherever there is auto-affection, wherever there is life, there is hetero-affection, impure auto-affection, and that impurity means, in a word, death. Wherever there is a self-relation, the two sides are split between activity and passivity, and yet the two sides are the same. In other words, and Derrida has shown this repeatedly, within auto-affection, such as hearing oneself speak, there is mediation, but mediation understood as "spacing," *espacement*. The self-relation is a relation that is at once joined and disjoined, contact and syncopated.[6] Derrida's conception implies that the self is related to a "me," a "who," which must be absolutely singular if it is to be genuinely *one* "me," and, at the same time, due to the mediating spacing, the "me" is repeated and universalized, a "what," which makes that "me" be not the self and be other than the self. This difference occurs in the very *moment* of auto-affection. Before we go any farther, we must note that in German the word for "moment" is *Augenblick*, literally, "the blink of an eye," *un clin d'œil*, and this blink, which closes the eye, means that the auto-affective moment includes blindness.

As you probably know, blindness has been a pervasive theme (if we can use this word) for Derrida throughout his career.[7] But, so far as I know, the words *blind* or *blindness* (*aveugle* or *aveuglement*) appear in the titles of his texts only once. This text is, of course, *Memoirs of the Blind* (*Mémoires d'aveugle*), an essay he wrote for the catalogue of an exhibition he organized at the Louvre in 1990.[8] It is this Derrida text that I would like to examine here.[9] *Memoirs of the Blind* is clearly related to *Voice and Phenomenon* — to the blink of the eye — and thus to the deconstruction of the metaphysics of presence. But *Memoirs of the Blind* is clearly different from *Voice and Phenomenon*. Here, as seems to be characteristic of this period in his writing, Derrida gives a noticeable privilege to *Christianity*, to the conversion of Saint Paul and the confessions of Saint Augustine. It seems to me that *Memoirs of the Blind* is more than just a phase in Derrida's deconstruction of the metaphysics of presence. Instead, it opens a larger, more ambitious project that we can call "the deconstruction of Christianity." Derrida had in fact anticipated this "wider" project at the end of "Violence and Metaphysics," where he speaks of the "jewgreek."[10] We might say that this deconstruction "transcends" any other.[11] In any case, my thesis is: in *Memoirs of the Blind*, Derrida is engaged in a deconstruction of Christianity.[12] The deconstruction takes place through the self-portrait, which is a figure of auto-affection. But this figure of auto-affection seems to center on one painting: Jan Provost's *Sacred Allegory*. The central point in what Derrida writes about this painting is that the eye has nothing to do with sight but with weeping, with tears.[13] What is important about tears is that they are ex-orbitant, they pass over the limit of the ocular globe. Tears of mourning for a departed lover, as they run down the face, draw lines, tracings, or figures, whose forms, we might say, proclaim, like Scripture (of course, *écriture*, in French), an other still to come, becoming thereby tears of joy. In my comments about *Memoirs of the Blind*, I will, of course, come to the tears. But I will also come to animals. At the end of *Memoirs of the Blind*, Derrida wonders whether animals can cry too. And if they can cry, do they not also, like us, wait for an other still to come, do they not also wait for the revelation? In any case, when we think about crying, we must think of the eye as a source-point, as an origin. So we will begin with what Derrida, in *Memoirs of the Blind*, calls "the origin of drawing."[14]

The Origin of Drawing

Almost immediately in *Memoirs of the Blind*, Derrida tells us that he is concerned with "the origin of drawing" (MdA 10/3). For Derrida, the origin of drawing is blindness. What does this statement mean? While the exhibition consisted primarily of drawings, the first picture seen in the exhibition was a painting by Joseph Benoît Suvée, "Butades or the Origin of Drawing."[15] It depicts a woman—a drafts-woman, in fact—Butades, drawing the features (*les traits*, as the French description says) of her lover, who is about to go away, by "tracing" his shadow on a wall (MdA 137/133).[16] Butades is able to make this "shadow writing" only by means of what we could call "un écart infime," "a minuscule hiatus."[17] Indeed, we could say that the origin of drawing (and therefore also of painting) is this minuscule hiatus. It is a kind of faultline that necessitates—and Derrida con-stantly stresses this connection, to which we shall return, between *une faute* and *il faut*.[18] So it is necessary that, *il faut que*, anyone who draws move back and forth between the vision of the model and the vision of the surface on which he or she is tracing. In between these two visions, there is the minuscule hiatus. Without this minuscule hiatus, one would *either* have the vision of the model *or* the vision of the paper, but *not* drawing on the paper. Yet this minuscule hiatus, without which there could be no drawing, indeed, without which there would be no duration, is unrepresentable (MdA 46/41). It is a moment of blindness, *un clin d'œil*. The eyes close, or the eye is gouged out, and that means, as Derrida says, that "drawing is blind, if not the draftsman or draftswoman" (MdA 10/2).

In *Memoirs of the Blind*, Derrida calls this structure of blindness in drawing "transcendental" (MdA 46/41). Now, while Derrida has used the word *écart* to designate this gap—a word that was much in vogue in France during the 1960s—he also, of course, uses the word *trait* (which we encountered already in the French description of the Suvée painting). Already in *Of Grammatology*, in 1967, Derrida had recognized that drawing consists in *le trait*—that is, the trait, feature, line, or stroke (these are all possible English translations of this ambiguous French term; see MdA 56n52/51n52). The "trait" or drawing establishes a formal difference, and Derrida calls this establishment *espacement*.[19] As he says in *Memoirs of the Blind*, "the trait spaces [*espace*] by delimiting"; it "joins and adjoins only by separating" (MdA 58/54). But "the tracing power of the trait" (*la puissance tracante du trait*), "the graphic act," is itself invisible, or "a-perspectival" (MdA 48–50/45). Playing on the word *perspective*, with its literal connection to sight (*aspicere*), Derrida calls this *a-perspectivism* the first *aspect* of what gives the experience of the *gaze* over to blindness (MdA 48/44–45), the first "powerlessness" (*impouvoir*) for the eye (MdA 48/44). But we should note here that, with this first structural aspect of the blindness of sight, it is impossible in Derrida to speak of a perspectivism. One cannot take up a perspective on the trait, since it establishes the difference that allows perspectives to be taken up. At the end, I intend to return to this Derridean a-perspectivism. In any case, in relation to this first powerlessness of the eye, Derrida says, "Whether it is improvised or not, the invention of the trait does not follow [*suit*], it does not conform to what is presently visible, to what would be posited in front of me as a theme" (MdA 50/45). What the invention of the trait separates is formal universality and informal singularity.[20] This separation between singularities and universals allows for this a-perspectivism to be interpreted in two ways.

On the one hand, the invisibility of the graphic act can be interpreted as memory, as *anamnesis* (MdA 50–56/45–51). In this case, drawing does not originate in perception. As we have seen, moving back and forth, the draftsperson cannot look at a model while drawing; therefore, when the draftsperson is drawing, the model, the other, is absent, perhaps, we would have to say, dead.[21] Derrida says that "from the outset [*dès l'origine*], perception belongs to recollection" (MdA 54/51). Yet this memory, which sees less, allows the draftsperson to see more, to be a visionary seeing beyond the present. Always in the past or always in the future, the form that the draftsperson invents is beyond any theme that could be posited.[22] But, *on the other hand*, the invisibility of the graphic act can be interpreted as *amnesia* (MdA 56–58 / 51–53, also CIR 34 / 33). Derrida says, "The visible *as such* would be invisible, not as visib*ility*, the *phenomenality* or *essence* of the visible, but as the *singular* [my emphasis; all others are those of Derrida] body of the visible itself" (MdA 56 / 51–52). What must be forgotten are the forms. In this forgetfulness, one must strain one's eyes to see exactly the one singular thing right there, naked. Here, with *amnesia*, we move from formal universality to informal singularity (or to unicity, *unicité*). Where above the memory of a form as universality was a theme, in the technical sense of something posited over and against me, below the forgetfulness attached to informal singularity will be a theme. But the forms cannot be forgotten. The singularity remains covered up, veiled, invisible, due to the forms. A respectful distance, tact, is maintained before the nude.

If the graphic act must remain invisible, so must the line itself when it is drawn. The invisibility of the line, once traced, is what Derrida calls the second powerlessness of the gaze. The trait withdraws. In a drawing, the colored thickness that is still attached to a line tends to become lighter and lighter in order to mark the sole border of a contour. When the line reaches this limit, there is nothing more to see. This "nothing more to see," Derrida says, is "the line itself" (MdA 58 / 54). As it approaches this threshold, the line makes only the surroundings of the trait appear, and then the trait itself is nothing (MdA 58 / 54). In other words, it can have no attributes, and therefore it requires a specific way of speaking, which brings us to the third and final powerlessness of the eye. Because the trait withdraws, there must be a "rhetoric of the trait" (MdA 58 / 56). We are following, of course, a line that runs from drawing to writing. Because there is no present model for drawing, the identification of it is always questionable, which gives rise to the title, to the "discursive murmur" (MdA 60 / 56). As Derrida says, "drawing comes in place of the name, which comes in place of drawing" (MdA 60 / 57). There is always therefore an "internal duel" between the title and the drawing. This internal duel or dual, the repetition or re-treat of the trait (*le retrait du trait*) brings us to the self-portrait, *l'auto-portrait*.[23]

The Logic of the Self-Portrait

It brings us to the self-portrait because the "transcendental retreat" "calls for" its own self-portrait. Since the origin of drawing is absent or invisible, it "calls for" a form or figure, a drawing or duplicate, in order to keep it in memory. And yet, as we know from our consideration of the logic of this origin, the transcendental retreat also "forbids" its own self-portrait; since the origin is singular, it cannot be reproduced. Because of this "calling for" and "forbidding," what is at issue in Derrida's considerations of the self-portrait is the certainty of the *identification* of the self portrayed (the *auto* portrayed). The exhibition contained, for example, certain self-portraits by Henri Fantin-Latour in which he is facing out (*vu de face*) and drawing, as if he were facing a mirror and drawing himself drawing the self-portrait. Derrida wonders whether we can be certain that the draftsperson is not "in the process of drawing something else [*autre chose*]" (MdA 64/60). In other words, can we be certain that the self in the portrait is Fantin-Latour drawing Fantin-Latour? Or, more precisely, can we be certain that the origin of drawing is singular or multiple? Of course, we are in the domain, as Derrida points out, of hypothesis, conjecture, and faith.[24] According to Derrida, this conjecture about the identification of the self that is portrayed in the picture comes about only under a certain condition, a spatial condition. He says, "In order to form the hypothesis of the self-portrait of the draftsman as self-portraitist, *and seen from the front* [*vu de face*], we, as spectators or interpreters, must imagine that the draftsman is staring at one point, at one point only, the focal point of a mirror that is facing him; he is staring, therefore, from the place that *we* occupy, in a face to face with him" (MdA 64/60, Derrida's emphasis).[25] Derrida also calls this point of a mirror "the source-point" (MdA 61/57), and we will return later to this "source" or "spring," this hole from which water flows. The logic of this point, however, consists in the following: in order for the draftsman to see himself as an object, he must position himself as a spectator; he must see himself as another would see him. Spacing must be established; there must be a hiatus—this is the *écart infime* about which we were just speaking—between himself and the mirror. This distancing opens up another place, and it opens the possibility of re-placements. As Derrida says, "if there were such a thing, the self-portrait would first consist in assigning, thus in describing, a place to the spectator, to the visitor" (MdA 64–65/62). The place of the mirror's single point therefore is where "we" are, all the virtual spectators; "we" or the others have replaced the mirror. Setting up a "circle of exchange" (MdA 96/94), this replacement, for Derrida, has several necessary implications.

First, since virtual spectators have replaced the mirror—the mirror is gone and we are there—the draftsman cannot see himself. The "signatory" or the "author," as Derrida says, has been struck blind. But second, "du même coup," again as Derrida says, because of the même, because of the "same," the model is struck blind. The self-portrait of the now blind draftsman is a self- or auto-portrait. If the draftsman is blind, and that is what is being represented, then the model is blind too. What is being drawn is blind; blindness becomes the model. In other words, "the subject" of the work, at once model, signatory or author, and object of the work (or sitter in the picture),[26] gouges out *his own eyes* in order to see himself as others would see him, with the others' eyes. And if this self-enucleation is necessary for the self-portrait, and if the draftsman intends to represent *himself*, then he must be represented as blind. For Derrida, however, there are still more implications. Third, as soon as we are "instituted" as spectators in place of the mirror, we are blinded to "the author *as such*" (MdA 65 / 62, my emphasis). The author requires necessarily the hiatus in order to see himself, but the hiatus means that he does not see himself as such, but as others see him. Thus he represents himself according to the others, *as other* (MdA 69 / 65). We can therefore no longer be certain about the identity of the drawing. Is Fantin-Latour drawing himself drawing or drawing himself drawing something else? This "something else," this *autre chose*, brings us to a fourth but not final implication. Even if it is clear that the self-portrait is a self-portrait of the draftsman drawing something else, some object other than the draftsman drawing himself, that object cannot be identified as such. It too must be an object for virtual spectators. As Derrida says, "there is no object, *as such*, without a supposed spectator" (MdA 66 / 63, my emphasis). The singular "as such" never appears, never appears naked; it is covered over, with spectacles, a mask, or a visor perhaps,[27] and forgotten. But in order to see the object *as something else*, there is, as Derrida says, "a bottomless indebtedness" to the other spectators, to a memory, that is, of the others. But, how many others? "To infinity," which only exacerbates the problem of identification. The inability to identify with certainty the self of the self-portrait—and this fifth implication is the last—implies the necessity of naming, the necessity of the title or the verbal event. As Derrida says, "Like Memoirs, the Self-Portrait always appears in the reverberation of several voices" (MdA 68 / 64).

As the subtitle to *Memoirs of the Blind* indicates, what we have here in this exacting "logic of the self-portrait" is ruin.[28] Ruin is a structural, logical, or transcendental feature of the work (MdA 72/68, 69/65). The failure of identification, the fault, the blindness is not an accident that supervenes (*survient*) on the work. Yet we know that blindness is not *just* a structure; the ruin of the eyes, their being used up, can happen as an event, and Derrida calls this event of blindness "the sacrificial event" (MdA 46/41). But Derrida insists that the transcendental and the sacrificial are never pure; there is always a hesitation between them (MdA 96/92).[29] This hesitation means that drawings of blindness, of men struck down with blindness as an event, as historical events of blindness, drawings with dates, in determinate cultures, are *also* drawings of the transcendental structure of blindness. The drawings then are stories or histories (*les histoires*) of the eye, of blindness, but also "pre-histories," which refer to the transcendental structure of blindness at the origin, at the beginning. But the event seems, for Derrida, to add something more. The violence inflicted on the eyes gives rise to *narratives*, narrative such as those of revelation and messianism (MdA 96/92, 46/41). While Derrida divides the violence inflicted on the eyes into three kinds—ruse, which always involves falling down, punishment, and conversion (MdA 96/92)—his entire discussion converges on the idea of *conversion*, which brings us to the third section of this chapter.[30]

The Conversion

The conversion of which Derrida is speaking happens when one is struck blind; then, even if one has fallen down due to a ruse and one is being punished, one is converted from corporeal vision to clairvoyance or providence (MdA 96/92). In order to refer to this movement from sight through blindness to clairvoyance or providence, Derrida uses terms and phrases that we all know from him: *usure*, with its double sense of using up and usury, and *plus de*, with its double sense of more and no more. But Derrida's discussions of conversion center on one, Saint Paul's conversion (MdA 116–17/112). You recall that Paul, on the road to Damascus, is struck blind and falls down from his horse, his blindness being a punishment for persecuting the Christians, and then he converts to Christianity. We know this story, since Paul himself recounts it twice in the Acts of the Apostles. But when Paul recounts this story, he is in fact *confessing* before a tribunal, confessing that he has converted from persecution of Christians to being a witness to the Christian faith (MdA 118/116). Paul's confession, of course, leads us to Augustine's *Confessions*. Here, in *Mem-*

oirs of the Blind, Derrida takes us into the heart of Augustine's *Confessions,* book 10, where Augustine "conjures away the temptations of sight and calls for this conversion from the light to the light" (MdA 119 / 117). In his confession, Augustine raises up "'invisible eyes' against the 'concupiscentia oculorum.'"[31] In particular, Derrida points out that Augustine indicts works of art, and especially painting, as tempting this desire of the eyes (MdA 120 / 119). But *Memoirs of the Blind* is a book about works of art. So Derrida asks, "Would Saint Augustine thus condemn the temptations of all Christian painting?" Derrida responds to this question by saying, "Not at all, just so long as a conversion saves it" (MdA 121 / 119). This salvation of painting through conversion brings us to Jan Provost's painting *Sacred Allegory,* which, as Derrida notes, is also called *Christian Allegory* (MdA 121n95 / 121n95).

Let me briefly remind you of the composition of this painting. Provost's *Sacred Allegory* is arranged vertically, with one eye at the top, the divine eye, and another eye at the bottom, the human eye. Two hands emerge from the human eye at the bottom. On the horizontal line, we have Jesus on the left and Mary on the right. In the middle, there is a blue globe. In the upper left-hand corner, there is the lamb of salvation or the sacrificial lamb. And, finally, in the upper right-hand corner, there is a book that is partially open; the book has seven seals hanging off it, which indicate that it is the Book of Revelation.[32] It seems to me that what Derrida has written about the painting's symbolism consists in three interrelated points.

First, the painting is a self-portrait of drawing. It "shows," as Derrida says, "the exchange of glances which makes the painting possible" (MdA 121 / 121). While we must return to Derrida's frequent use of the verb *devenir,* we must recognize that the first point means that the painting "becomes" this allegory. Or, as Derrida says, "It stages the opening of the sacred painting, an allegorical self-presentation of this 'order of the gaze' to which a Christian drawing must submit" (MdA 121 / 121). As a self- or auto-presentation, the painting is a tautology (*auto*: same; cf. MdA 10 / 2). The painting represents a mirroring relation between divine vision and human vision; there is a relation of resemblance and analogy (MdA 121 / 119). The painting therefore "ordains" (*ordonne*) corporeal vision for divine vision. Yet, since the painting is an allegory (*allos*: other), there is, at the same time as there is tautology, heterology.[33] There is "infinite distance," the outside, and difference. In other words, the painting refers to "the law of disproportion, dissymmetry, and expropriation"

(MdA 121 / 121). Derrida can claim that there is a reference to alterity in the *Sacred Allegory* because of the allusion to the Book of Revelation in the painting. He says, "This [painting] shows an apocalypse, as the allusion to the Apocalypse of John—to the book 'sealed with seven seals'—suggests" (MdA 123 / 121). An apocalypse is a revelation, a making naked, the unveiling of truth; but an apocalypse is also an event, a singular event, which has not yet happened, an event that will overturn everything. Indeed, it will be a catastrophe or cataclysm, which ruins the order between human vision and divine vision (MdA 123 / 122). Again, as Derrida says, "A work is at once order and its ruin"; it "shows its origin, the condition of its possibility, and the coming [*venue*] of its event" (MdA 123 / 122). Derrida's use of the *venue* here brings us to the second point in what he has written about Provost's *Sacred Allegory*.

So, second, *Sacred Allegory* is a self-portrait of a *conversion*, and here, of course, it is important to hear the literal meaning of the word *conversion*, a turning toward (*vers*). But we also need to hear the active sense of the word; conversion is a process of becoming (*un devenir*), a "convert*ing*." What is being converted in this painting is what Augustine had called the *concupiscentia oculorum*. In other words, what is being converted is the "scopic impulse," voyeurism, the "desire" of the eyes (MdA 72 / 68, 121 / 121). It is the conversion of the love of corporeal vision into the love of the light of God. In the process of converting, the love of corporeal vision would be in the process of becoming a love of a singular Being, about whom no attribution could be made and, at the same time, becoming a love of a universal form, about which one can indeed make attributions. If love is being converted in this way, it would be in the process of becoming the love of an "impossible totality" (MdA 72 / 68).[34] Love would be in the process of becoming the love of the self, of a "one," a "who," united, naked, intact, joined to the self without "being out of joint," and yet this self would always be "out of joint," untotalizable, divided by the "what," covered over, tactful, and plural. Repeatedly, in the discussion of this second point, we have been using the phrase "in the process of"; this phrase translates the French *en cours de*. Derrida says that the gaze must "be in the process of converting [*en cours de conversion*]." But he also says that "The gaze must become Christian [*le regard doit devenir Chrétien*]" (MdA 121 / 121). The *en cours de* and the *devenir* mean that something is happening here, something is being done here. This "doing" brings us to the third point that I think Derrida is making in what he has written about *Sacred Allegory*.

In the painting, the desire of the human eye is being called back, is being made to "recall [*rappelle*]" (MdA 121 / 121). The desire of the eyes being recalled to God (or to the ahuman) — Derrida says that "this [being recalled] is memory itself" — means that the painting is also a kind of memoir or a confession. This is our third point: insofar as it is a self-portrait, the painting is a confession. Now, as Derrida says, "in our culture," that is, "in Christian culture, there is no self-portrait without confession" (MdA 119 / 117). Again, Derrida refers to Augustine's *Confessions*, where Augustine repeatedly says that, since God knows everything in advance, confessions do not teach anything to Him. Derrida concludes from this that: "The self-portraitist . . . *does not lead one to knowledge*, he confesses [*avoue*] a fault [*une faute*] and asks for forgiveness. He 'makes' [*fait*] the truth, to use Augustine's word, *he makes* [fait] *the light* of this narrative, throws light on it, in order to make the love of God grow within him, 'for love of your love'" (MdA 119 / 117, Derrida's emphasis).

In *Sacred Allegory*, as I already pointed out, two hands emerge from the human eye at the bottom of the picture. Hands, of course, do things or make things; *ils font*. For Derrida, the hands emerging from the eye at the bottom of the picture are imploring, "an imploration of surrection and resurrection" (MdA 123 / 121). That is, they are hands uplifted from a position of falling, of fallenness. In other words, they are confessing. Or, as Derrida says, they are hands in the process of joining together for prayer and hymn, prayer being a speaking to God, to a "who," and hymn being a speaking about God, about a "what."[35] And we should not forget what Derrida stresses in *Circumfession*; there is a connection between *prier* and *prendre*.[36] By praying, the hands are asking for something to be given that they can take, they are asking for for-giveness. But, we should not be speaking of speaking here. The prayer is silent; the hands are for writing; writing is what they do. But, most importantly, imploring from the position of fallenness and fault, the hands that are flowing out of the eye are like tears. The "source-point," about which we were speaking earlier, is not the position of the eye seeing. The source-point has become a spring from which water flows or courses. The desire of the eyes, the scopic impulse, must always be *en cours de*, that is, coursing and flowing with tears. And tears streaming down the face make tracks, tracings, almost a kind of writing. By "writing," these tears do more than see; they make something happen or come (*fait arriver, fait venir*), an event. We know already what they are "writing" about: the Apocalypse.

Conclusion: "For the Creation Waits with Eager Longing for the Revelation"

In *Memoirs of the Blind*, Derrida says that what he is writing about Provost's *Sacred Allegory* is *"not* an analysis" (MdA 121 / 121, my emphasis).[37] For Derrida, an "analysis is always more and something other than analysis. It transforms; it translates a transformation already in progress."[38] In short, what Derrida has written about *Sacred Allegory* is a deconstruction. Now it seems to me that we could say that what Derrida has written about *Sacred Allegory*, indeed *Memoirs of the Blind* as a whole, is a more current phase in the deconstruction of the metaphysics of presence. It consists in overturning the intuitive and immediate presence of vision into sight veiled with tears. As Derrida had already done in *Voice and Phenomenon*, in *Memoirs of the Blind* he shows the essential impossibility of eliminating mediation. Using the old phenomenological terms, we could say that he shows again the irreducibility of *Vergegenwärtigung* in every *Gegenwärtigung*. And this irreducibility of *Vergegenwärtigung*—we could say that the structure of *Vergegenwärtigung* cannot itself be deconstructed—implies that Derrida has also replaced the "primacy of perception" with the "primacy of memory." But *Memoirs of the Blind* seems to be something more than a deconstruction of the metaphysics of presence. As we have just seen, Derrida gives a striking privilege to Christianity in this book. For Derrida, Paul's confessions—or his memoirs—are more than just any memoirs. Derrida says, "Let us wager that Paul's confession . . . will have come to represent the model of the self-portrait, the model of the one that concerns us here in its very ruin" (MdA 119 / 117). This comment means that Paul's conversion to Christianity is the very model of everything Derrida has been speaking of in *Memoirs of the Blind*. Therefore, I think one has to recognize that *Memoirs of the Blind* (and perhaps other texts of this period, such as "How to Avoid Speaking") opens up what we must characterize as a project that is larger, more ambitious than a deconstruction of the metaphysics of presence. It opens up the "wider" project of a deconstruction of Christianity.[39]

It is possible to claim, as Jean-Luc Nancy has in 1998, that, because Christianity replaces the Greek, the Jewish, and the Roman, its scope is larger than any one of these. Thus a deconstruction of Christianity is larger than a deconstruction of the metaphysics of presence, since the "metaphysics of presence" is only Greek.[40] In his article on the deconstruction of Christianity—which Derrida mentions in *Le Toucher*[41]—Nancy stresses that the deconstruction of Christianity would aim to designate, within Christianity, an *archē*, a beginning, a "provenance" of Christianity (and therefore of the West) that is "more profound that Christianity itself."[42] In other words, it would be a provenance that is no longer Christian.[43] If the origin of Christianity is no longer Christian, then I think we must also recognize that Derrida privileges Christianity in *Memoirs of the Blind* in order to give a representation and a name—a proper name, perhaps—to the unrepresentable and the unnameable, to the transcendental structure and the empirical event, to the source-point itself and as such (cf. MdA 58 / 54).[44] Or, as in his 1984 essay on Paul Celan, "Shibboleth," this name could be "Judaism." Still again, as Derrida suggests in his 1996 "A Word of Welcome" (on Levinas), this name could be "Sinai," which, being a frontier, implies that the provenance is neither Jewish nor Christian nor Islamic.[45] There is always "dis-identification" and "expropriation" in the source-point, which dislocates the name, making it improper, opening it to the stranger.[46] As dis-identifiable and expropriatable, as unrepresentable and unnameable, this source-point says nothing, and therefore this "religion" is a kind of atheism.[47] Here, we have a god who has said "Adieu."[48] Therefore, immediately after speaking of Augustine at the end of *Memoirs of the Blind*, Derrida turns to Nietzsche, the anti-Christ, whose word is "God is dead." Yet as Derrida points out, Nietzsche too cried a lot. If the deconstruction of Christianity follows a path that intersects with Nietzsche—and Nancy says that it does through the question of nihilism[49]—then the deconstruction of Christianity is going to have to concern the body, *le corps propre* or *la chair*, the flesh, as we say in English, or, as is said in the German phenomenological tradition, *die Leib*.[50] But the path to the body proper leads to the question of life. Perhaps life is another name for, another example of, the unrepresentable and the unnameable.

It seems to me that, at this point in the history of philosophy, perhaps of our culture, we require a new concept of life. It is impossible to invent a new conception of life unless one confronts Heidegger. In his lecture courses on Nietzsche, given during the 1930s, Heidegger criticizes Nietzsche's concept of life as will to power.[51] He shows that at the center of "the metaphysics of the will to power" lies valuation (*Wertsetzung*). Indeed, he says that "Valuation [in Nietzsche] is the fundamental occurrence of life itself."[52] Now, Heidegger connects valuation back to Nietzsche's well-known perspectivism.[53] In "Nietzsche's Word 'God is Dead,'" a speech Heidegger gave in 1943 based on the final years of the courses, he says: "The essence of value lies in its being a point of view [*Gesichtpunkt*]. Value means that upon which the eye is fixed."[54] According to Heidegger, that upon which the eye is fixed is values, understood as conditions for the preservation and enhancement of power. Because in Nietzsche, according to Heidegger, the eye looks for, that is, values, *only* what can preserve and increase power (super-abundant life), because the eye never seems to let beings be, in "Nietzsche's Word 'God is Dead,'" Heidegger says, quite dramatically, I think, that "thinking in terms of values is radical killing" and "the value-thinking of the metaphysics of the will to power is murderous in a most extreme sense."[55]

Now, it also seems to me that only what Derrida has done in *Memoirs of the Blind* provides a way to move along the path from the deconstruction of Christianity to life; it provides the only way to move beyond Heidegger.[56] The origin of drawing that Derrida lays out here, as we have seen, is an a-perspectivism. Derrida has found the inability (*l'impouvoir*) in the midst of power, the blindness in the midst of vision.[57] Here, in contrast to the value thinking that Heidegger attributes to Nietzsche, we would have Derrida's thought of the unconditional gift, which, if it is worthy of its name, "would not appear *as such* to the donor or donee."[58] The gift would be an event of blindness. So, in contrast to the perspectivism that Heidegger attributes to Nietzsche, Derrida says, in *Memoirs of the Blind*, that "the best point of view (and the point of view will have been our theme) is a sourcepoint and a water-point—which thus comes down to tears" (MdA 128/126).[59] The conversion that strikes blind—Paul's conversion, for example—converts the desire of the eyes, the appetite. Insofar as this desire now moves toward the "impossible totality," insofar as it now remembers, it "renounces its own impulse, its own movement of appropriation."[60] And yet, in its powerlessness, the desire of the eyes opens out onto a different kind of power or ability, which is weeping, praying, writing. This "doing" (*faire*) calls all of us back from the murderous thinking of values.

We must note one last thing about this murderous thinking of values. Still in "Nietzsche's Word 'God is Dead,'" Heidegger tells us that this thinking, this form of subjectivity, has led to man's being an "insurrection" (*Aufstand*) among all the beings, among all creation, turning everything into an object of technology.[61] Man becomes the master of nature and of the animal. Now, according to Derrida in "The Animal That Therefore I Am," this insurrection of man over the animals "is based paradoxically on a fault and a failing in man."[62]

Earlier, we have seen this essentially necessary fault in our discussion of the origin of drawing. But in "The Animal That Therefore I Am," Derrida stresses that this fault is a default in the very propriety of man. The default in propriety implies that man "can never possess the pure, rigorous, indivisible concept as such of [the] attribution" of properties that man attributes to himself in order to rise up over the animals.[63] No property is ever exclusively that of man, even the tears that seem to be proper only to man's eyes, according to, as Derrida tells us in *Memoirs of the Blind*, "the anthropo-theological discourse" of Christianity (MdA 128 / 126).[64] This fault in man implies that we can say that animals weep. We know the story about Nietzsche; before he fell into dementia, he wept for those poor horses in Turin (MdA 125 / 126).[65] But can we say that horses weep for man, for themselves, and all the animals? In *Memoirs of the Blind*, writing of Caravaggio's famous painting *The Conversion of Saint Paul*, Derrida says, "only the horse remains standing. Lying outstretched on the ground, eyes closed, arms open and reaching up toward the sky, Paul is turned toward [*vers*] the light that bowled him over. The brightness seems to fall upon him as if it were reflected by the animal itself [*sa bête même*]" (MdA 117 / 112). Perhaps we can say that the light flowing back out of the horse is like tears. Perhaps Paul's horse therefore is imploring and confessing, even asking for forgiveness. Then, like ours, the suffering of animals calls everyone back from the murderous thinking of values. Like us, they too are in need of salvation. And in this regard, we need to recall what Paul wrote to the Romans (8:19): "For the creation [and the creatures] waits with eager longing for the revelation."[66] Their waiting, perhaps, would mean that the animals not only react but also respond.

4

Eschatology and Positivism
The Critique of Phenomenology in Derrida and Foucault

In his early "What Is Metaphysics?" Heidegger claims that the question expressed in the title of his essay puts the questioner — *us* — in question. This "putting us in question" then moves toward what Heidegger terms the completion of the transformation of man, understood as subject, into existence (*Dasein*).[1] This complete transformation, for Heidegger, as we know from the introduction that he added to the essay in 1949, amounts to an overcoming (*Überwindung*) of metaphysics understood as Platonism or as the mere reversal of Platonism (WM 363/279). At this moment, I think it is still necessary to take seriously Heidegger's attempt to overcome metaphysics.[2] Heidegger had pointed the way toward the overcoming of metaphysics by calling us to think what he calls the *Auseinander* of the opening of being itself (WM 369/284). How are we to translate into English this German word *Auseinander*? Perhaps as the "outside of one another" or even as the "outside itself." No matter what, however, *Auseinander* implies that, in order to overcome metaphysics, we must have a thought of the outside. A thought of the outside would be a thought that, coming from the outside, is equally a thought about the outside. This outside, it seems to me, is not "the opening of being," as Heidegger says, but the opening of life. The outside is a place in which life and death indefinitely delimit one another. But to move us to this place of delimitation, we must start with a critique of phenomenology.

We must start here because phenomenology has shown a remarkable resilience across the twentieth century. More importantly, we must start here because phenomenology has already conceived life through its central concept of *Erlebnis*, "lived-experience," or *vécu*. Therefore, we can ask whether phenomenology itself has already initiated an overcoming of metaphysics. Husserl, of course, thought so. Yet certain critiques in France dating from the 1960s imply that lived-experience consists in a kind of insideness that is not internal and a kind of sameness that is not identity but mixture and ambiguity. If mixture and ambiguity define lived-experience, then it follows that *sometimes* phenomenology restores Platonism, *while at other times* it merely reverses Platonism into its opposite. Understood in this way as sameness and insideness, phenomenology does not overcome metaphysics. Phenomenology is not a thought of the outside—or, at least, that is what I will seek to show. The allusion in the phrase "the thought of the outside" is, of course, to Foucault, in particular, to his critique of phenomenology in *Words and Things*. The other critique comes from Derrida's *Voice and Phenomenon*. What I intend to do here is reconstruct the critique of phenomenology found in Foucault and Derrida.[3] I will start with Foucault, in particular, with chapter 9 of *Words and Things*: "Man and His Doubles."

The Analysis of Lived-Experience (*Vécu*) Is a Discourse with a Mixed Nature

"Man and His Doubles" contains, of course, Foucault's critique of modern humanism.[4] The chapter therefore focuses on man (and not on the human being). Foucault defines man as a double; he is at once an object of knowledge and a subject that knows (MC 323/312). Man (again, not the human being) is what occupies, as Foucault says, this "ambiguous position." The entire critique of humanism unfolds, for Foucault, from this designation of man as "ambiguous," a designation that recalls, of course, Merleau-Ponty and Sartre. I shall turn to Merleau-Ponty in a moment. In any case, for Foucault, the ambiguity that defines man consists in two senses of finitude. In one sense, finitude consists in the empirical positivities, the empirical contents of "work, life, and language," which tell man that he is finite (MC 326/315). The knowledge of life, for instance, tells man that he is going to die. The other sense is that this finitude is itself fundamental. The forms of knowledge whose very contents tell man that he is finite are themselves finite. For man, there is no intellectual intuition, for instance. So finitude is ambiguous between empirical content and foundational forms. For Foucault, this ambiguity of finitude results in an "obligation" to ascend "up to an analytic of finitude." Here it is necessary to hear the word *analytic* in its Kantian sense, as a "theory of the subject" (MC 330/310). For Foucault, this would be an analytic "where the being of man will be able to found, in their positivity, all the forms that indicate to him that he is *not* infinite" (MC 326/315, my emphasis). This analytic would be the discourse of phenomenology.

The discourse of phenomenology aims, according to Foucault, at a truth that would be neither empirical content nor transcendental form, while trying to keep the empirical and transcendental separated. This is an important qualification, since what is at issue is whether phenomenology can maintain the separation between the empirical and the transcendental. In any case, according to Foucault, phenomenology would be an analytic of man as a subject in this precise sense: man as subject, "that is, as the place of empirical knowledge but led back as close as possible to what makes empirical knowledge possible, *and* as the pure form that is immediately present to these contents." Man as subject therefore would be a third and intermediary term in which empiricity and transcendentality would have their roots. According to Foucault, this third and intermediary term has been designated *le vécu*. *Le vécu* responds to the "obligation" to analyze finitude, that is, to the obligation to have a theory of the subject. Here is Foucault's definition of *le vécu*: "lived-experience, in fact, is at once the space where all empirical content is given to experience; it is also the originary form that makes empirical content in general possible." We can now see the problem with *le vécu*, indeed, with "man." *Le vécu* must be concrete enough so that one could apply to it a descriptive language; yet it must be sufficiently removed from positivity so that it can provide the foundations for empirical positivity. The discourse of *vécu* tries to make the empirical hold for the transcendental: the empirical is the transcendental and the transcendental is the empirical, or, the content is the form and the form is the content. Lived-experience therefore is a mixture. Thus Foucault says that "the analysis of lived-experience [*vécu*] is a discourse with a mixed nature: it is addressed to a specific but ambiguous layer" (MC 332/321). This analytic "mixes" the transcendental and the empirical together. Therefore the concept of lived-experience, as Foucault understands it—and this is also how Derrida understands it—consists not in an identity of empirical content and foundational forms but in a mixture or ambiguity between these two. Here, however, one could plausibly wonder whether such a definition can be found in phenomenology. So let us turn now to Husserl and then to Merleau-Ponty to confirm this definition.

Lived-Experience (*Erlebnis, le vécu*) in General

We have been discussing *Erlebnis*; let us turn to Husserl's classic definition of *Erlebnis* in *Ideas I*, Section 36, "Intentional Lived-Experiences: Lived-Experiences in General."[5] In order to distinguish what he is doing from psychology, Husserl says, "Rather [than a discourse of *real* psychological facts; the word *real* is, of course, important] the discourse here and throughout is about purely phenomenological lived-experiences, that is, their essences, and on that basis, what is 'a priori' enclosed in [*in beschlossen*] their essences with unconditional necessity" (Hua III:1, p. 80).[6] It is important that Husserl calls psychological facts "real," because all purely phenomenological lived-experiences are *reelle*. What Husserl calls intentional lived-experiences, thoughts in the broadest sense, are *reelle*, which means that thoughts are internal. Yet intentional lived-experience also contains "the fundamental characteristic of intentionality," the property of being consciousness of something. This "of something"—the fundamental characteristic of intentionality—means that lived-experience is related to an outside; something comes from the outside into lived-experience. But, Husserl says, "within the concrete unity of an intentional lived-experience," there are *reelle* moments, which do not have the fundamental characteristic of intentionality; these *reelle* moments are the data of sensation. Here, Husserl has discovered something nonintentional and therefore passive at the very heart of lived-experience, something that comes from the outside, and yet he has designated these moments as *reelle*, and thereby as "enclosed in" *das Erlebnis überhaupt*. By means of this *überhaupt* and this *in beschlossen*, we can conclude already that *Erlebnis*, in this classic formulation, consists in a sameness that is not identity and an insideness that is not simply internal; in a word, *Erlebnis* "in general" consists in a mixture.

To demonstrate this sameness and insideness again, I would like to look at another Husserl text: the final version of Husserl's 1927 *Encyclopedia Britannica* entry for "phenomenology." This text introduces phenomenology through phenomenological psychology, which, Husserl says, has the task of investigating the totality of lived-experience. More importantly, phenomenological psychology, according to Husserl, is an easier way to enter into the transcendental problem that occurred historically with Descartes, namely, that all of reality, and finally the whole world, are for us in existence and in existence in a certain way only as the content of our own representations. Thus everything real has to be related back to us. But this "us" cannot be the psyche, according to Husserl, because the psyche is defined by the mundane sense of being as *Vorhandenheit*, "presence," or, more literally, "presence-at-hand." To use a mundane being — whose ontological sense is *Vorhandenheit* — to account for the reality of the world (whose ontological sense is also *Vorhandenheit*) is circular, and this circularity defines psychologism.[7] In contrast to psychologism, phenomenology claims, according to Husserl, that there is a parallelism between psychological subjectivity and transcendental subjectivity and that this parallelism involves a deceptive appearance (*Schein*) of "transcendental duplication." It is important here, it seems to me, that, while Husserl recognizes that there is some sort of difference between the transcendental and the psychological or the empirical, he does not, we might say, partition off the transcendental from the psychological or empirical. Instead, he says that transcendental subjectivity is defined by *Vorhandenheit*, too, but "not in the same sense [*nicht im selben Sinn*]."[8] Indeed, Husserl thinks that by saying "not in the same sense" he has eliminated the deceptive appearance and makes the parallelism understandable. He says, "the parallelism of the transcendental and the psychological spheres of experience has become comprehensible . . . as a kind of identity of the interpenetration [*Ineinander*] of ontological senses."[9] He also describes this "kind of identity" as "ambiguity" (*Zweideutigkeit*). Here Husserl thinks the *Ineinander*, literally, "one in the other," but not, we might say, the *Auseinander*, literally, "one outside of the other." Nevertheless, this *Zweideutigkeit* and *Ineinander* should make us think of Merleau-Ponty. So I would like to turn now to Merleau-Ponty, in particular, to his *Phenomenology of Perception*.[10]

On the very first page of *Phenomenology of Perception*, Merleau-Ponty speaks of *vécu*, and throughout the *Phenomenology* the word modifies the word *monde*, "world." In the chapter called "The Phenomenal Field," for example, Merleau-Ponty says that "the first philosophical act therefore would be that of returning to the lived-world on this side of the objective world" (PhP 69/57). Yet he uses the word as a noun — *le vécu* — only twice. The first time occurs in the chapter called "Space," where he says "lived-experience [*le vécu*] is really lived by me . . . , but I can live more things than I can think of [*plus de choses que je m'en représente*]. What is only lived is *ambivalent*" (PhP 343/296; my emphasis). For Merleau-Ponty, ambivalence is the crucial characteristic of *vécu*. And this characteristic guides his analysis of intersubjectivity in *Phenomenology of Perception*, which is where he uses *le vécu* for the second time, in the chapter called "Others and the Human World." There *le vécu* is defined by self-givenness (PhP 411/358), but this self-givenness is also given (PhP 413/360). In other words, the active is also passive. In this formula we can see the importance of the positive affirmation in the *is*. This positive affirmation is the heart of ambivalence. Now these two uses of *le vécu* in *Phenomenology of Perception* depend, of course, on Merleau-Ponty's appropriation of Husserl's concept of *Fundierung*. In the chapter called "The Cogito," Merleau-Ponty speaks of the relation between founding (*le fondant*) and founded (*le fondé*) as one that is "equivocal" (*équivoque*), since "every truth of fact *is* a truth of reason, every truth of reason *is* a truth of fact" (PhP 451/394; my emphasis).[11] Merleau-Ponty also says that the relation between matter and form is a relation of *Fundierung*: "The form integrates the content to the point that it appears to end up being a simple mode of the form . . . but reciprocally . . . the content remains as a radical contingency, as the first establishment or the foundation of knowledge and action. . . . It is this dialectic of form and content that we have to restore" (PhP 147–48/127). We can now summarize what we see in Merleau-Ponty's concept of *le vécu*. For Merleau-Ponty, *le vécu* is ambivalent or equivocal—it is, we could say, a mixture, *un mélange*—because the content of experience, *le sol*, as Merleau-Ponty also says, becomes, is integrated into, the form of expression. Phenomenological lived-experience therefore is not defined by identity but by sameness and mixture of form and content, or of empirical and transcendental.

"Un écart infime, mais invincible"

Both Husserl and the early Merleau-Ponty conceive *Erlebnis* as mixture and ambiguity because both want to overcome the duality of subject and object, or even the duality of what Heidegger calls the ontological difference. In other words, phenomenology is an attempt to overcome Platonism or Cartesianism (dualisms) by mixing together content and form. In both Foucault and Derrida, we find statements asserting that the phenomenological concept of *Erlebnis* mixes in this way. First we have Foucault's statement in chapter 7 of

Words and Things, called "The Limits of Representation." Foucault says:

> Undoubtedly, it is not possible to give empirical contents transcendental value, or to move them onto the side of a constituting subjectivity, without giving rise, at least silently, to an anthropology, that is, to a mode of thought in which the in principle limits of knowledge [*connaissance*] are *at the same time* [*en même temps*] the concrete forms of existence, precisely as they are given in that same empirical knowledge [*savoir*]. (MC 261 / 248, my emphasis)[12]

Even if phenomenology is transcendental, Foucault is saying, it still falls prey to a "silent anthropology." It takes *my* present or *our* present experiences, which are content, as foundational forms. In other words, on the basis of the empirical contents given to *me*, or, better, to *us*, phenomenology tries to determine the form of that empirical content. While trying to keep them separate, phenomenology makes the transcendental and the empirical the same. It confuses them (MC 352 / 341). Now, in the Introduction to *Voice and Phenomenon*, Derrida makes a very similar statement, but he adds something that helps us see the principle of the critique: "Presence has always been and will always be, to infinity, the form in which—we can say this apodictically—the infinite diversity of content will be produced. The opposition—which inaugurates metaphysics—between form and matter finds in the concrete ideality of the living present its ultimate and radical justification" (VP 5 / 6). When Derrida says here that the opposition between form and content finds its ultimate and radical justification, he means that content, the root of empirical positivity, and form, the finality of transcendental foundation, are mixed together in the living present *at the same time*. Indeed, in both quotes we see that the mixture of subject and object in lived-experience depends on a temporal sameness: "at the same time" or "simultaneity," *en même temps* or *à la fois*. This dependence on temporal sameness tells us already that a critique of the concept of lived-experience will come from a kind of spatial thinking and from a reinstitution of dualisms.

We can see the critique most clearly in Derrida, in chapter 6 of *Voice and Phenomenon*, "The Voice That Keeps Silent." Derrida's critique there centers on the concept of presence. Here is the definition of presence he provides: "presence [is] *simultaneously* [à la fois] . . . *the being-before of the object*, available for a look and . . . *proximity to self in interiority*. The 'pre' of the *present ob*ject now-before is an *against* [contre] (*Gegen*wart, *Gegen*stand) simultaneously [à la fois] in the sense of the *wholly against* [tout-contre] of proximity and in the sense of the *encounter* [l'encontre] of the op-posed" (VP 83–84/75; Derrida's emphasis). Presence, as Derrida understands it, is *à la fois* close by and proximate, and *à la fois* far off and distant. In other words, it must be "at the same time" self-presence and presence, the object as repeatable to infinity and the presence of the constituting acts to themselves. For Derrida, this ambiguity between presence of an object and self-presence of a subject is found in the voice of interior monologue, in other words, hearing oneself speak. The primary characteristic of this "absolutely unique type of auto-affection" (VP 88/78) is temporality. When I speak to myself silently, the sound is iterated across moments. This temporal iteration is why, as Derrida explains, sound is the most ideal of all signs (VP 86/77). Thus, in hearing oneself speak, one still exteriorizes one's thoughts or "meaning-intention" or acts of repetition in the iterated and iterable phonemes. This exteriorization—ex-pression—seems to imply that we have now moved from time to space. But, since the sound is heard by the subject during the time he is speaking, the voice is in absolute proximity to its speaker, "within the absolute proximity of its present" (VP 85/76), "absolutely close to me" (VP 87/77). The subject lets himself be affected by the phoneme (that is, he hears his own sounds, his own voice, *la voix propre*) without any detour through exteriority or through the world, or, as Derrida says, without any detour through "the non-proper in general" (VP 88/78). Hearing oneself speak is "lived [*vécue*] as absolutely pure auto-affection" (VP 89/79). What makes it be a pure auto-affection, according to Derrida, is that it is "a self-proximity which would be nothing other than the absolute reduction of space in general" (VP 89/79). Yet—and this is a crucial "yet"—there is a double here between hearing and speaking. As Derrida says, this pure auto-affection, which is the very root of transcendental *Erlebnis*, supposes that "a pure difference . . . divides the presence to oneself" (VP 92/82). This difference divides the *auto*. As Derrida says, "It produces the same as the self-relation within the difference from oneself, the same as the nonidentical" (VP 92/82). Being nonidentical, auto-affection is ambiguous. We must understand the nonidentity, however, in the following way: when I hear myself speak, the hearing is a repetition of the speaking that has already disappeared; re-presentation (*Vergegenwärtigung*) has intervened, and that intervention means, in a word, space. As Derrida says, "the 'outside' insinuates itself in the movement by which the

inside of non-space, what has the name of 'time,' appears to itself, constitutes itself, 'presents' itself" (VP 96/86). Within time, there is a fundamental "spacing" (*espacement*) (VP 96/86).[13] Derrida also calls this spacing *un écart* within *le vécu* (VP 77/69). On the basis of Derrida's use of the word *écart*, we can rejoin Foucault.

In *Words and Things*, Foucault says that all of the doubles in which man consists are based on "un écart infime, mais invincible," "a hiatus, minuscule and yet invincible" (MC 351/340). Here we can dissociate an ambiguity in the word *infime*. This *écart* is *infime*, that is, *minuscule*, insofar as it is minuscule, the *écart* closes and relates "in the manner of 'a mixed nature.'" But this *écart* is also *infime* in the sense of infinitesimal, infinitely divisible, and thus a great distance that separates and keeps open. It seems to me that this *écart infime* sets up all our problems. In fact, I think it is impossible to overestimate the importance of "Man and his Doubles." Foucault says there, after mentioning this minuscule hiatus, that, in contrast to classical thought, in which time founds space:

> in modern thought, what is revealed at the foundation of the history of things and of the historicity proper to man is the distance hollowing out the Same, it is the hiatus [*écart*] that disperses the Same and gathers it back at the two edges of itself. It is this profound spatiality that allows modern thought still to think time—to know it as succession, to promise it as completion, origin, or return. (MC 351/340)

It seems to me, if I may extend the analysis a bit, that we must see this "profound spatiality" working, as well, in Deleuze's critique of phenomenology, found both in his 1968 *Difference and Repetition* and in his 1969 *The Logic of Sense*.[14] For Deleuze, the phenomenological concept of *Urdoxa*, which one finds both in Husserl and in Merleau-Ponty, is not originary, since it is always "copied off"—*décalqué*—*doxa*, or common sense.[15] This "copying off" means that the *Urdoxa* is mixed with or the same as the *doxa*; they resemble one another and are not differentiated. The phenomenological concept of *Urdoxa* has violated, therefore, the most basic principle of Deleuze's thought, perhaps the most basic principle of thought itself: "The foundation can never resemble the founded." Deleuze continues, "It is not enough to say about the foundation that it is another history—it is also another *geography*, without being another world."[16] For Deleuze, the earth is a profound spatiality, consisting in "un écart infime, mais invincible."

Conclusion: Positivism and Eschatology

The critique of phenomenology found in Foucault and Derrida, and in Deleuze, is based in this minuscule hiatus. Despite the fact that all three — Derrida, Deleuze, and Foucault — share the same critique, there is a difference between them. To conclude, I am going to outline the difference between Derrida and Foucault. For both, the critique of phenomenological lived-experience is a critique of auto-affection. The critique depends entirely on one necessary possibility: wherever there is sensing, it must be possible for there to be a surface, and wherever there is a surface, it must be possible for there to be space. This necessary possibility implies that auto-affection, being alone and therefore close to oneself and unified with oneself, is always already virtually double, distant from oneself and divided. *But* — this is an important "but" — what divides the *auto*, spacing it and making it double *in Derrida* is mediation, *Vergegenwärtigung*. Derrida always conceives the *écart infime* through *Vergegenwärtigung*, representation. In Derrida, re-presentation contaminates presentation; mediation, in other words, contaminates the immediate, *but contamination is still mediation.* Thus, understood as mediation, contamination promises *unity*, even though it cannot, of necessity, ever keep this promise. The other is always already close by and coming, without ever arriving. Without ever being able to arrive, the one who is going to keep the promise is to come in person (in the flesh, *Leiblich*). Therefore, we must characterize Derrida's critique of phenomenology (as he himself has done) as an *eschatological* critique. It is a critique based in a promised unity that demands to be done over again and again.[17]

Like that of Derrida, Foucault's critique depends entirely on one necessary possibility: wherever there is sensing, it must be possible for there to be a surface, and wherever there is a surface, it must be possible for there to be space. This necessary possibility implies that auto-affection, being alone and therefore close to oneself and unified with oneself, is always already virtually double, distant from oneself and divided. *But* — this is where we see the difference from Derrida — what divides the *auto*, spacing it and making it double *in Foucault* is a battle.[18] Foucault conceives the *écart infime* as a battle. The opponents in the battle are words and things, or hearing and seeing. The battle consists in attacks and crossings across the surface (*entrecroisements*), *but these attacks do not form a unity.* No unity is ever *promised* in the battle. Politics, which looks to be peace, as Foucault points out in

Discipline and Punish, is war being fought by other means. For Foucault, the audio-visual battle is an immediate relation. There is no mediation because the opponents can never be mixed together or, we might say, can never contaminate one another. Instead, the opponents are posited as such; there is always the opposition of *resistance*. Therefore, we must characterize Foucault's critique of phenomenology (as he himself has done) as a *positivistic* critique.[19] It is a critique based in a duality without negation and thus it is entirely positive.[20]

What are we to make of the difference in their critiques? We must return once more to "Man and His Doubles." Here, Foucault lays out a kind of genealogy of phenomenology. At the beginning of the nineteenth century, he tells us, there was a dissociation, in the double sense of finitude, between empirical content and foundational forms of knowledge. This dissociation was Kant's thought. The dissociation, however, led to what Foucault calls a transcendental aesthetics (the empirical content) and a transcendental dialectic (the foundational forms). The transcendental aesthetics became positivism; the transcendental dialectic became eschatology. During the nineteenth century and at the beginning of the twentieth century, this dissociation between positivism and eschatology came to be associated, in two ways: Marxism and phenomenology. We can see the association in Marxism insofar as Marxism claimed to give the positive truth of man in conditions of labor and *at the same time* promised a revolutionary utopia. We can see this association in phenomenology insofar as phenomenology speaks of the content of *Erlebnis*, which can be positively described as the truth, and *at the same time* of the fulfillment of a meaning-intention, in other words, the promise of fulfilled truth. For Foucault, this association leads to the ambiguity that defines both Marxism and phenomenology. It seems to me that, in their similar but different critiques of phenomenology, Foucault and Derrida have once again dissociated positivism and eschatology. The association that phenomenology and Marxism made has become unraveled. The doubles that came to be the ambiguity of Husserl's thought, positivism and eschatology, have now themselves become dissociated into the thought of Foucault and Derrida. On the one hand, we have Derrida's messianism, which leaps back to the eschatology of the nineteenth century. On the other hand, we have Foucault's "fortunate positivism" (*un positivisme heureux*),[21] which obviously leaps back to the positivism of the nineteenth century. Foucault and Derrida have dissociated immanence and transcendence, faith and knowledge, and, we might even say, the heart and the brain. Both the brain and the heart are complicated spaces; we might even appropriate Heidegger's term *Auseinander* in order to conceive them. Yet without the heart one could not speak of life, and without the brain one could not speak of memory. Now we can see what to make of the difference between the critiques of phenomenology that we find in Foucault and Derrida. This is our task. We must continue the overcoming of metaphysics by trying to find a new way of associating the heart and the brain. In other words, is it possible for us to find a new distribution, a new *partage*, between the double of the heart and the brain?

5

Un écart infime (Part I)
Foucault's Critique of the Concept of Lived-Experience (Vécu)

In 1984, at the end of his life, Foucault revised the introduction he had written in 1978 for the English translation of Georges Canguilhem's *The Normal and the Pathological*. Foucault gave no title to the original introduction, but in 1984 he gave it the simple title "Life: Experience and Science."[1] Here, Foucault tried to show that Canguilhem "wants to re-discover . . . what of the concept is *in life*" (VES 773–74 / 475; Foucault's emphasis). For Canguilhem, but also for Foucault himself, we must think that the concept is immanent *in — dans* —life.[2] What is at issue in immanence is the logic of this relation between concept and life. Clearly, one could just as well say that phenomenology consists in the immanence of the concept in life. Yet just as clearly, Foucault thinks that what Canguilhem was doing with the concept of life was radically different from the phenomenological concept of life. In fact, this is what Foucault says at the end of his revised introduction: "It is to this philosophy of sense, of the subject, of lived-experience [*le vécu*] that Canguilhem has opposed a philosophy of error, of the concept, of the living [*le vivant*] as another way of approaching the notion of life" (VES 776 / 477). Here I intend to examine this difference between *le vécu* ("lived-experience") and *le vivant* ("the living"), that is, I intend to examine the different logics, we might say, of immanence that each concept implies. To do this, I am going to reconstruct the "critique" that Foucault presents of the concept of *vécu* in chapter 9 of *Words and Things*, "Man and

His Doubles."[3] Then I am going to construct the positive logic of Foucault's relation of immanence by means of another text, contemporaneous with *Words and Things*: *This Is Not a Pipe*.[4] As we will see, the critique of the concept of *vécu* is based on the fact that the relationship in *vécu* is a mixture (*un mélange*), which closes *un écart infime*. Conversely, Foucault's conception of the relationship—here we must use the word *vivant*—in *le vivant* is one that dissociates and keeps *l'écart infime* open. Perhaps I will give my conclusion away if I say that for Deleuze—whom we must also keep in mind here—immanence is defined by a kind of dualism, a dualism that "is a preparatory distribution within a pluralism," within, in other words, a multiplicity.[5]

Lived-Experience (*le Vécu*) in Merleau-Ponty

In chapter 9, Foucault names no particular philosopher when he criticizes the concept of *vécu*. But we know from "Life: Experience and Science" that, for Foucault, the side of the subject and *le vécu* refers to phenomenology, more particularly, to Sartre and Merleau-Ponty. Thus it is probable that Foucault, in chapter 9, is thinking of the early Merleau-Ponty, the Merleau-Ponty of the *Phenomenology of Perception*. Foucault's use of the word *écart*, to which we shall return, also makes us think of the Merleau-Ponty of *The Visible and the Invisible*. Below, I shall turn to the later Merleau-Ponty. But here at the beginning, we shall remain with the Merleau-Ponty of the *Phenomenology of Perception*.[6] On the very first page of the *Phenomenology of Perception*, Merleau-Ponty speaks of *le vécu*, and throughout the *Phenomenology* the word modifies the word *monde*, "world." In the chapter called "The Phenomenal Field," for example, Merleau-Ponty says that "the first philosophical act therefore would be that of returning to the lived-world on this side of the objective world" (PhP 69 / 57).[7] Yet he uses the word as a noun—*le vécu*—only once (so far as I know). In the chapter called "Space," he says "lived-experience [*le vécu*] is really lived by me . . . , but I can live more things that I can think of [*plus de choses que je m'en représente*]. What is only lived is *ambivalent*" (PhP 343 / 296; my emphasis). For Merleau-Ponty, ambivalence is the crucial characteristic of *vécu*. And this characteristic guides his analysis of intersubjectivity in the *Phenomenology of Perception* (PhP 411 / 358). Here "lived solipsism" (*solipsisme vécu*) is defined by self-givenness (PhP 411 / 358), but this self-givenness is also given (PhP 413 / 360). In other words, the active is also passive. In this formula we can see the importance of the positive affirmation in the "is." This positive affirmation is the heart of ambivalence. These two uses of *le vécu* depend, of course, on Merleau-Ponty's appropriation of Husserl's concept of *Fundierung*.[8] In the chapter called "The Cogito," Merleau-Ponty speaks of the relation between founding (*le fondant*) and founded (*le fondé*) as one that is "equivocal" (*équivoque*), since "every truth of fact *is* a truth of reason, every truth of reason *is* a truth of fact" (PhP 451 / 394; my emphasis).[9] Merleau-Ponty also says that the relation of matter and form is a relation of *Fundierung*: "The form integrates the content to the point that it appears to end up being a simple mode of the form . . . but reciprocally . . . the content remains as a radical contingency, as the first establishment or the foundation of knowledge and action. . . . It is this dialectic of form and content that we have to restore" (PhP 147–48 / 127). We can now summarize what we see in Merleau-Ponty's concept of *le vécu*. For Merleau-Ponty, *le vécu* is ambivalent or equivocal—it is, we could say, a mixture, *un mélange*—because the content of experience, *le sol*, as Merleau-Ponty also says, becomes, is integrated into, the form of expression. This relation would have to be formulated as a positive affirmation; the copula indicates the sameness of things related. We know, however, that the logic of the *Fundierung* relation in Merleau-Ponty is not yet complete. Since he calls it a dialectic, it must involve some sort of negation. We shall return to the question of negation in a moment. Now let us turn to Foucault's "critique" of the concept of *vécu* in *Words and Things*.

The Analysis of Lived-Experience (*Vécu*) Is a Discourse of a Mixed Nature

As we have seen, "Man and His Doubles" contains Foucault's critique of modern humanism. Foucault defines man, of course, as a double; he is at once an object of knowledge and a subject that knows (MC 323 / 312). As doubled, man is ambiguous. The entire critique of humanism unfolds, for Foucault, from this designation of man as "ambiguous," a designation that recalls Merleau-Ponty (but perhaps not Sartre, at least not the Sartre that Merleau-Ponty portrays in *Adventures of the Dialectic*). We have seen that, for Foucault, the ambiguity consists in two senses of finitude. In one sense, finitude consists in the empirical positivities, the empirical contents of "work, life, and language," which tell man that he is finite (MC 326 / 315). "The knowledge of life" (Canguilhem), for instance, tells man that he is going to die. The other sense is that this finitude is itself fundamental. The forms of knowledge whose very contents tell man that he is finite are themselves finite. As Kant showed, for instance, for man there is no intellectual intuition. So finitude is ambiguous between empirical content and foundational forms. For Foucault, this ambiguity of finitude results in an "obligation" to ascend "up to an analytic of finitude." Here is it necessary to hear the word *analytic* in its Kantian sense, as a "theory of the subject" (MC 330 / 310). For Foucault, this would be an analytic "where the being of man will be able to found, in their positivity, all the forms that indicate to him that he is not *in*finite" (MC 326 / 315).

For Foucault, because the analytic of finitude consists in "bringing to light the conditions of knowledge on the basis of the empirical contents which are given in the knowledge" (MC 329 / 319),[10] two kinds of analyses arise in the nineteenth century. In both of these analyses, Foucault has Marxism in mind. On the one hand, there is what Foucault calls a "transcendental aesthetics," in which one discovers that "knowledge had anatomo-physiological conditions"; this transcendental aesthetics would be "a nature of human knowledge." On the other hand, there is what Foucault calls a "transcendental dialectic," in which one would study "the illusions that are more or less ancient, more or less difficult to eliminate, of humanity"; this would be "a history of human knowledge." Here we can see that the two senses of finitude have been dissociated between a "positivism"—this is the transcendental aesthetics—and an "eschatology," which is the transcendental dialectic. This dissociation calls, as Foucault says, for "a critique," in the sense of providing the conditions for the possibility of positivism and eschatology. Without a critique, positivism and eschatology remain naïve. This critique is *a distribution of the truth*. In particular, what is required is "a truth which would allow us to have, concerning the nature or history of human knowledge, a language that would be true." In other words, what is required is a discourse that would be neither of the order of a reduction to positive truth or of the order of a promise of truth revealed. This discourse is that of phenomenology.

The discourse of phenomenology would aim at both requirements, while trying to keep the empirical and transcendental separated. It would be an analytic of man as a subject in this precise sense: man as subject, "that is, as the place of empirical knowledge but led back as close as possible to what makes empirical knowledge possible, *and* as the pure form that is immediately present to these contents." Man as subject therefore would be the third and intermediary term in which positivism and eschatology would have their roots. According to Foucault, this third and intermediary term has been designated *le vécu*. *Le vécu* responds to the "obligation" to analyze finitude, that is, to the obligation to have a theory of the subject. Here is Foucault's definition of *le vécu*: "lived-experience, in fact, is at once the space where all empirical content is given to experience; it is also the originary form that makes the empirical / content in general possible." We can now see the problem with *le vécu*, indeed, with "man." *Le vécu* must be concrete enough so that a descriptive language can be applied to it, yet it must be sufficiently removed from positivity to provide its foundation. The discourse of *vécu* still tries to make the empirical hold for the transcendental. A simple judgment of equivalence could express this "hold for," this kind of immanence: the empirical is the transcendental and the transcendental is the empirical, or, the content is the form and the form is the content. We have returned to Merleau-Ponty's equivocity: *le mélange*. And thus Foucault says that "the analysis of lived-experience [*vécu*] is a discourse with a mixed nature: it is addressed to a specific but ambiguous layer" (MC 332 / 321). This analytic "mixes" the transcendental and the empirical together in an affirmative judgment. But this affirmation brings us to the question of negation.

In chapter 9, Foucault does not explicitly speak of negation. But in a second discussion of phenomenology, he recognizes that phenomenology, being a reflective philosophy, transforms the old idea of thought thinking itself into thought thinking its other. This other is called "the unthought" (MC 337 / 326). The word *unthought* (*l'impensé*) obviously contains a negative prefix. This is what Foucault says about "the unthought": "it has never been reflected upon for itself according to an autonomous mode . . . it has received the complementary form and the inverse name" (MC 337–38 / 327). This citation means that "the unthought" or "the unconscious," for instance, has never received its own positive and autonomous form; it has always been that which is *not* thought or that which is *not* consciousness. This negation would even mean that "the unthought" is that which is devoid of the form of thought, that which is emptiness itself, and therefore that about which one can say nothing. As early as *The History of Madness*, in 1961, Foucault had discovered this structure of negation. Unreason (*déraison*) is the experience of madness (*la folie*) as the lack of truth, "the non-being of error," and as the lack of reason, "the empty negativity of reason." Foucault calls *déraison* "the night," "the obscure contents [tied up] with the forms of clearness." The result is that "all of what madness can say about itself is nothing but reason."[11] Again, nothing positive can be said about "the unthought" or "the unconscious" or "unreason." Being emptiness, the irrational is nothing, but as soon as we speak of it, we give it the form of reason, which implies that it is nothing but reason. Now *The History of Madness* can help us understand the negation in one other way. Here we move from form to content. Because unreason (*déraison*) is emptiness (*le vide*), it can be filled with the content of reason. Thus, for Foucault, we say that unreason is *not* reason because reason "is taken in an aberrant face."[12] This aberration is "an extreme, negative slenderness" (*une extrême minceur négative*), "a negative index," which means that unreason is nothing but "quasi-reason." Foucault calls this "negative index" *un écart*.[13] This *écart* brings us back to *Words and Things*.

For Foucault, all of the doubles in which man consists are based on "un écart infime, mais invincible"; the English translation says, a "hiatus, minuscule and yet invincible" (MC 351 / 340). Here we can dissociate an ambiguity in the word *infime*. This *écart* is *infime*, that is, *minuscule*; insofar as it is minuscule, the *écart* closes and relates in the manner of "a mixed nature." Here, the *écart* has the sense of a deviation from a norm. But, this *écart* is also *infime* in the sense of infinitesimal, infinitely divisible, and thus a great distance that separates and keeps open. This sense of the negative word *in-fime* clarifies one of the most infamous things that Foucault says in chapter 9. In the section called "The Empirical and the Transcendental," he advises that, if one wants to "contest" both positivism and eschatology truly—in other words, if one want truly to construct a critique—one should try to imagine that man does not exist. This "paradox" means: try to imagine a theory of the subject different from the modern theory of the subject (man). This different theory of the subject would be an analytic, too, but now in the literal sense of the word *analytic*, in the sense of loosening, of untying, of taking apart, even of differentiating within a *mélange*. *This* analytic would not "mix" the transcendental and the empirical together but would make their difference *infime*.

This Is Not a Pipe

In order to clarify the infinitesimal sense of this *différence infime*, we will now turn to Foucault's analysis of a painting by Magritte called

This Is Not a Pipe. The analysis of this painting is at least analogous, if not identical, to that of *le vécu* in *Words and Things*. Just as lived-experience is a *mélange* of the empirical and the transcendental, this picture looks to be a calligram. A calligram is literally "beautiful writing," words drawn in figures. Magritte's picture consists in a drawing of a pipe floating in air above a sentence that says "ceci n'est pas une pipe." In other words, we have a figure and a text that names it. Foucault calls it a "calligram" because the picture looks to be written and the text looks to be drawn. According to Foucault, a calligram has a triple function: "to compensate the alphabet; to repeat without the help of rhetoric; to capture things in the trap of a double cipher" (CP 20 / 20–21). This quote means first that a calligram makes the figure speak and the words represent. We can see here the old oppositions between "showing and naming; figuring and saying; reproducing and articulating; imitating and signifying; looking and reading" (CP 22 / 21). But these oppositions are now effaced, because the text and the figure are, as Foucault says, "tautological" (CP 21 / 21). For Foucault, although the text and the figure are the same, this sameness does not mean that Magritte's picture is an allegory of something else, of something that is somewhere else; without any allegory, the calligram repeats without the help of rhetoric. Instead, the calligram attempts to trap the thing itself. In order to spring this trap, the calligram makes use of a particular property of letters. "At once," letters have the value of "linear elements that we can arrange in space *and* . . . signs that we must unfold according to the unique chain of the sonorous substance" (CP 21–22 / 21, my emphasis). Thus twice the calligram tracks the thing itself. Pure discourse cannot represent the thing; pure drawing cannot say the thing. The calligram, by contrast, at once draws and says the thing itself. Here we can see in Foucault's description of the calligram's triple function the same structure as we saw in *le vécu*. We have a double between saying and figuring, or between the empirical and the transcendental, or even between life and concept, but this double is really the same, tautological (*to auto*). Even more, here we have a same that tightly closes this small, infinitesimal distance between the two.

Yet Magritte's picture is not a simple calligram. According to Foucault, here, in fact, we have a different logic of the relation between. For Foucault, Magritte has not only constituted a calligram but has also "carefully unmade" it (*défait avec soin*) (CP 19 / 20). This is why it produces in us "an indefinite uneasiness." Magritte has perverted the triple function of the calligram. First, instead of the words invading the figure and vice versa, in Magritte's picture the words have returned to their old place at the bottom of the page; they have become a legend. Yet because the words look to be drawn and the figure looks to be written, Magritte, according to Foucault, has distributed words and things in their traditional disposition "only in appearance" (CP 24 / 22). But there is no tautology here between words and things, since, on the one hand, we have a figure that is so familiar that it has no need of being named, and, on the other, at the very moment that the legend should give us the name it gives us the name by denying it: "ceci n'est pas une pipe." Here we start to see the importance of negation, of the negative adverb, in Foucault's thought. For a calligram to function, it is necessary that the viewer (*le voyeur*) look (*regarde*) and not read. Then the picture (which is made out of words) is a pipe. But as soon as the viewer becomes a reader, then the picture is no longer a pipe but a sentence with a sense. Indeed, it is not a pipe but a sentence. According to Foucault, Magritte has understood that "the calligram never says and represents, either by ruse or impotence—it hardly matters—at the same moment" (CP 28 / 24–25). In other words, because of the way in which Magritte has distributed the space, separating the picture from the words, the picture says that "I am a thing (or a pipe or a picture of a pipe) and I am not words," while the sentence says "I am not a pipe and I am words" (CP 29 / 25). We no longer have a tautology here. Instead, as Foucault says, "the redundancy of the calligram is based on a relation of exclusion; the hiatus [*l'écart*] of the two elements in Magritte, the absence of letters in his drawing, the negation expressed in the text bring forward *affirmatively two positions*" (CP 29–30 / 25–26, my emphasis). We have two positions affirmed—and not a *mélange*—by means of the double negation. The *écart* remains here *infime*, in the sense of infinitesimal or indefinite.

But Foucault makes one more point; in fact, it is the essential one. Because we still have the remnants of a calligram in Magritte's painting, "it is therefore necessary to admit that between the figure and the text there is a series of crossings between [*entrecroisements*]" (CP 30 / 26). Foucault calls this series of crossings "a battle" (Deleuze, in fact, calls it "the audio-visual battle"[14]). For Foucault, this battle takes place through the "this," the *ceci*, through the index, which is the subject of the sentence. The subject of the sentence has become the space of a battlefield. This space of the battlefield is what we find when we try to imagine that man as a subject no longer exists. In fact, for Foucault what has happened is that the subject has been reduced to the infinitive of the verb. If we recall the literal meaning of the word *verb*, then we can even say that the subject has been reduced to the voice or, as Foucault would say, to the murmur. We would be able to see this reduction of the subject down to the infinitive if we had time to examine the paradox with which Foucault opens his essay on Blanchot, *The Thought from the Outside*.[15] From the title of this essay alone, we can see that the infinitive of the verb is how Foucault would think about thought itself, about the concept. The paradox of "I speak" (*je parle*) suspends the transitive function of language, leaving behind only the dispersion of the verb (*parler*), the dispersion of the voice: verbs of becoming. But we can say as well that the infinitive of the verb, which precedes declensions and tenses, which precedes, in other words, subjects and times, is like the trajectory of an arrow. According to a very old paradox, the trajectory of an arrow can be infinitely divided, implying that it has no beginning and no end.[16] This infinite movement could also be called error, the error by means of which one could start to think about what *le vivant* means and "another way of approaching the notion of life" (VES 775 / 477). This other way of approaching the notion of life would be multiplicity.

Three Landmarks in the Later Merleau-Ponty

To conclude, I would like to state, precisely as I understand it, Foucault's critique of the concept of *vécu*, in other words, his critique of phenomenology and therefore of the early Merleau-Ponty. Because of the movement of anti-Hegelianism that dominated French philosophy in the sixties, there was a kind of return to Kant. Thus Foucault's critique of phenomenology (which is somewhat different from that of Derrida) consists in showing that phenomenology's central concept of lived-experience, although post-Kantian, was based on a pre-Kantian naïveté.[17]

Phenomenology itself consisted in the attempt to overcome Neo-Kantianism by showing that the conditions of experience can themselves be experienced.[18] But this way of overcoming Neo-Kantianism resulted in confusing the difference between conditions and conditioned (MC 352 / 341).[19] In other words, phenomenology resulted in a sameness of conditions and conditioned. In *Words and Things*, Foucault recognizes that the phenomenological task Husserl set himself at the end of his life connects phenomenology, "in its deepest possibilities and impossibilities," to the historical destiny of Western philosophy as it was established in the nineteenth century. But, as we have seen, Foucault says, "Undoubtedly, it is not possible to give empirical contents transcendental value, or to move them onto the side of a constituting subjectivity, without giving rise, at least silently, to an anthropology, that is, to a mode of thought in which the in principle limits of knowledge [*connaissance*] are at the same time the concrete forms of existence, precisely as they are given in that same empirical knowledge [*savoir*]" (MC 261 / 248). Thus, even though phenomenology is transcendental and even if it revives the deepest intentions of Greek metaphysics, it still falls prey to a silent anthropology. Now, I think it is also significant that Foucault, in 1966, in *Words and Things*, targets not only phenomenology but also Marxism for being naïve. Of course, this critique will result in Foucault's new idea of political philosophy, presented in *Discipline and Punish* in 1974. In light of Foucault's double target—both phenomenology and Marxism—we should recall what Merleau-Ponty says in the final published working note to *The Visible and the Invisible*: "the conception of history that one will come to will be in no way like that of Sartre. It will be much closer to that of Marx."[20] I've just mentioned *The Visible and the Invisible*. Let us now quickly see whether Foucault's critique still strikes at the later Merleau-Ponty.[21] To do this, let me indicate only three "landmarks" in the complicated terrain of Merleau-Ponty's later philosophy. The first landmark is painting, more precisely, the paintings of Paul Klee.

In "Eye and Mind," Merleau-Ponty quotes Klee's tombstone, which says, "I cannot be grasped in immanence."[22] For Merleau-Ponty, this inscription means that painting cannot be grasped in phenomenological immanence,[23] which refers us once more to lived-experience, to *le vécu*. But as we saw in *Phenomenology of Perception*, Merleau-Ponty's concept of *vécu* is equivocal or ambivalent. In *Eye and Mind*, this ambivalence means that Klee's paintings, especially his reflections on line, "eat away at prosaic space and its *partes extra partes*" (OS 75 / 143). If space is not *partes extra partes*, parts outside of one another, then, it seems, space must be conceived as a kind of sameness. This sameness is why Merleau-Ponty also quotes Klee, saying that, in a forest, he sometimes feels not only that he is looking at the trees but also that the trees are looking back at him (OS 31 / 129). What we have in this space is a sameness of looking at and being looked at, or even a sameness of imitation and signification, of words and things, a kind of "narcissism" (VI 183 / 139).[24] In *This Is Not a Pipe*, Foucault describes Klee's project as overcoming the duality and hierarchy between resemblance and reference. Thus what is at issue in Klee's paintings, for Foucault, is not the calligram or the collage; "rather what is at issue is the intersection [*l'entrecroisement*] of the system of representation by resemblance and the system of reference by signs *in one same network*" (CP 42 / 33, my emphasis). Klee's paintings involve "crossings"; *entrecroisements* is the exact word Foucault had used to describe the audiovisual battle in Magritte. These crossings are why Foucault says that Klee's enterprise is "not foreign" to that of Magritte (CP 45 / 35). Yet for Foucault, Magritte's enterprise is "opposed" to that of Klee. Why are Klee and Magritte opposed? In Klee, the space of the crossings is not heterogeneous, but mixed together "in one same network." This one same network is why Foucault, in a 1966 interview, says that "in relation to our century [i.e, the twentieth], Klee represents what Velázquez had been able to represent in relation to his [that is, the seventeenth]."[25] In fact, in the discussion after his 1951 presentation of "Man and Adversity," Merleau-Ponty defines *ambiguity* precisely as a mixture in "one same."[26] This idea of mixture or ambiguity brings us to the second landmark.

The second landmark is the debate with Sartre.[27] As we know from *Adventures of the Dialectic*, Merleau-Ponty believed that Sartre's ontology is *not* genuinely dialectical.[28] Thus, in *The Visible and the Invisible*, in the chapter entitled "Interrogation and Dialectic," Merleau-Ponty claims that Sartre's ontology is a form of "surveying thought" (VI 129 / 87). This characterization means that Sartre has abstracted being and nothingness from concrete experience; he thereby transforms being and nothingness (or the negative) into abstract categories of thought. Merleau-Ponty's critique of Sartre, however, is not limited to this charge of abstraction. Because being and nothingness are separated from experience, Sartre is required to find a way of unifying being and nothingness, a way of unifying the in-itself and the for-itself, if he wants to overcome a dualistic ontology. Thus, Merleau-Ponty accuses Sartre of "constructing" a union between being and nothingness (VI 290 / 237), as if these categories were "ingredients" (VI 94 / 66). Merleau-Ponty calls this constructed union a *mélange* (VI 290 / 237). Despite Merleau-Ponty's association of the word *mélange* with Sartre, it is clear that the idea of a *mélange* is going in the right direction for Merleau-Ponty.[29] In saying, for instance, that Sartre's "philosophy of the negative . . . described our factual situation with more penetration than had ever before been done," Merleau-Ponty characterizes Sartre's penetrating description as a *mélange* (VI 120 / 87). And again, Merleau-Ponty says, based on his discussion of Sartre, that philosophy "is a question put to what does not speak . . . it addresses itself to that *mixture* [*mélange*] of the world and of ourselves that precedes reflection" (VI 138 / 102, my emphasis).[30] For Merleau-Ponty, this mixture also characterizes "our mute life," which brings me to the third and final landmark.

When Merleau-Ponty speaks of life in *The Visible and the Invisible*, still in the "Interrogation and Dialectic" chapter, he starts from "our life," which, by using the first-person plural viewpoint, confirms his phenomenological outlook. He then moves to the impersonal *on*, as the "one of corporeal life and the one of human life" (VI 116 / 84). This *on* or "one" is "the present and the past, as a pell-mell set of bodies and minds, promiscuity of faces, words, actions, with, between them all, that cohesion which cannot be denied them since they are all differences, extreme divergences [*écarts*] of *one same something*" (VI 116–17 / 84, my emphasis). The crucial phrase in this quote is the "one same something"; therefore, Merleau-Ponty defines life, both our life and the anonymous life of the "one," as an "inextricable implication" (VI 117 / 84).[31] This "implication," as Merleau-Ponty says, "cannot be torn apart by the accidents of my body, by death, or simply by my freedom" (VI 117 / 84, my emphasis). With this word *accident*, Merleau-Ponty has relativized death to life. What he is trying to do is *not* let death be an absolute limit. If it were an absolute limit, then he would fall into the abstract problems of being and nothingness that he found in Sartre: an absolute nothingness opposed to an absolute being would reduce the two to a simple identity. Thus, Merleau-Ponty says:

> The principle of principles here [it is not clear what Merleau-Ponty means with this *ici*: is he referring to the local discussion of negativity and Sartre or to the entire *Visible and the Invisible?*] is that one cannot judge the powers of life by those of death, nor define without arbitrariness life as the sum of forces that resist death, as if it were the necessary and sufficient definition of Being to be the suppression of non-being. (VI 117 / 84–85)

This quote means that, despite the accident of death, there is still, for Merleau-Ponty, the implication, in the literal sense, of you being in me and vice versa; there is still "one same something"; there is still, we might say, "synthesis."

In the passage that I just quoted, Merleau-Ponty is referring, of course, to the definition that Xavier Bichat gave to life in his 1800

Physiological Investigations on Life and Death: "life consists in the sum of functions that resist death."[32] *For Foucault*, this definition of life is one of Bichat's great innovations.[33] By defining life in this way, which means that life is the sum of functions that resist "degeneration," Bichat had relativized death. But death, for him, was no longer conceived as an accident.[34] Bichat, as Foucault says in *The Birth of the Clinic*, "volatized [death], distributed it throughout life in the form of separate, partial, progressive deaths, deaths that are so slow in occurring that they extend beyond death itself" (NC 147 / 144). For Foucault as for Merleau-Ponty, there is an *on*, a "one," but this "one" is the one of "one dies." But Bichat could make the discovery that life is the sum of functions that resist death only "from the viewpoint of death." What Foucault insists on is that only from the viewpoint of "decomposition" in its literal sense—and not from the viewpoint of "our life"—was Bichat able to give a "positive truth" to life. Death becomes then "the great analyst" of life, which "unfolds" and "unties the knot of life" (NC 147 / 144). In other words, death, understood as the slow process of decomposition, dissociates the different powers of life, just as Magritte's picture dissociated the powers of looking and reading. Instead of Merleau-Ponty's implication, with Foucault we have *ex*plication. In life, for Foucault, disease and resistance to disease are the opponents in the battle. As I already suggested, the other way of approaching the notion of life is through the concept of multiplicity. Only from the viewpoint of death can one see the multiple ways in which disease and decomposition, resistance and growth, battle one another. And only when one can see the multiplicity is it possible to reach the *singular*, the innumerable singularities that make up a multiplicity. Here, perhaps, we encounter the most striking difference between Foucault and Merleau-Ponty. For Foucault, it is not Husserlian nor Bergsonian lived-experience that allows us "to think living individuality," it is Bichat in his clinical experiments. As Foucault says in *The Birth of the Clinic*, "A century earlier [than Husserl and Bergson], Bichat gave a more *severe* [one has to hear in this word also "sever," "several," "separate," and "rigorous"] lesson. The old Aristotelian law, which prohibited the application of scientific discourse to the individual, was lifted when, in language, death found the place of its concept: space then opened up the differentiated form of the individual to the gaze" (NC 175 / 170, my emphasis).[35] Only from the viewpoint of death was it able and is it still able to write a case, and thus to think about "a life."[36]

6

Un écart infime (Part II)
Merleau-Ponty's "Mixturism"

Today, "immanence" and "transcendence" present a lot of problems. In 1959, Rudolf Boehm published an essay in which he claimed that these terms are already ambiguous in Husserl's work as early as 1907.[1] But perhaps we can impose some conceptual rigor on them. Let us say that immanence is opposed to the transcendent, meaning that the thought of immanence is the thought of life, while the thought of the transcendent refers to the old metaphysics, especially the two-world structure of Platonism. We can then distinguish within the concept of immanence two ways of being an anti-Platonist. On the one hand, one can say that ideas or concepts are immanent in life where the preposition *in* designates some sort of relation of *resemblance* between concept and life. On the other, one can say that ideas or concepts are immanent in life where the preposition *in* designates a relation of *incompatibility* between concept and life. Difference, *un écart*, defines both of these anti-Platonisms, but the first would be a thought of the same (which is not identity), while the second would be a thought of the dual (which is not a dualism of substances). As we have seen,[2] this sameness applies to the early Merleau-Ponty, the Merleau-Ponty of *Phenomenology of Perception*. In *Phenomenology of Perception*, Merleau-Ponty makes the concept of *vécu* ambiguous; lived-experience for him consists in a mixture. Foucault confirms this when, in chapter 9 of *Words and Things*, he says that "the analysis of lived-experience [*vécu*] is a discourse with *a mixed nature*"[3] (MC 332 /

321, my emphasis). So I have tried to show that at least the thought of the early Merleau-Ponty, the thought of the "good equivocity," as Merleau-Ponty would say, is a "mixturism."

Does this sameness, this "mixturism," apply to all of Merleau-Ponty's thought, to his later thought in particular as we see it presented in "Eye and Mind" (OE 87 / 148) and *The Visible and the Invisible* (VI 328 / 274–75)?[4] If we want to be conceptually rigorous, what I just called the thought of the same, which defines mixturism, is not, in fact, a thought of immanence, but rather a thought of transcendence. Here, I am making a distinction between transcendent and transcendence, where transcendence remains an anti-Platonism. But the main idea is that wherever there is resemblance, analogy, and equivocity, there is transcendence. This change in terminology from immanence to transcendence is more consistent with Merleau-Ponty's own usage, as we shall see. What I would like to show now is that we must conceive the relation of immanence as a relation between memory and life, in which the *and* in the phrase refers to vision. Vision is in the middle of memory and life. But in the middle of vision is an impotence of vision, a blind spot. We find the blind spot most clearly in Foucault's analysis, in *Words and Things*, of Velázquez's painting *Las Meninas*. Now, it is possible to argue that we can find a blind spot in Merleau-Ponty himself. It seems to me, however, that there is a subtle shift in emphasis from Merleau-Ponty to Foucault concerning blindness, a subtle shift that makes all the difference. But before we come to the shift in emphasis, let us start with Merleau-Ponty's mixturism.

The Conception of Merleau-Ponty's Mixturism

"Eye and Mind" is the last text Merleau-Ponty published during his lifetime. He wrote it during the summer of 1960 and published it in January 1961.[5] Immediately after the initial publication of "Eye and Mind," during the spring semester of 1961, at the Collège de France Merleau-Ponty was teaching a course called "Descartes's Ontology and Contemporary Ontology."[6] Following the structure of "Eye and Mind," but also expanding on it, the lectures fell into two parts: fundamental thought given in art, and then Descartes's ontology. The lectures on Descartes were given during April 1961, right up to Merleau-Ponty's death on May 3, 1961. At the beginning of the Descartes lectures, Merleau-Ponty says:

If Descartes's philosophy consists in this, [first, in the] establishment of a natural intelligible light against the sensual man [*l'homme sensuel*] and the visible world, then [second, in] the relative justification of feeling [*du sentiment*] by the natural light, it must contain . . . an ambiguous relation of light and feeling [*sentiment*], of the invisible and the visible, of the positive and the negative. It is this relation or this *mixture* [*ce mélange*] that it would be necessary to seek. (NdC 1959–61, 222, my emphasis)[7]

At the end of his life, Merleau-Ponty himself was seeking the mixture of the visible and the invisible. We can see the pursuit of mixturism already in Merleau-Ponty's 1947–48 lectures on the union of the body and the soul. In the second lecture, he says: "In Descartes, the question of the union of the soul and the body is not merely a speculative difficulty as is often assumed. For him, the problem is to account for a paradoxical fact: the existence of the human body. In the Sixth Meditation, the union is 'taught' to us through the sensation of hunger, thirst, etc, which issue from the mixture [*mélange*] of the mind with the body."[8] How are we to conceive Merleau-Ponty's mixture?

One conception of a mixture that we can rule out immediately is Sartre's dialectic of being and nothingness. According to Merleau-Ponty in *The Visible and the Invisible*, Sartre starts from abstract concepts of being and nothingness, that is, concepts abstracted from experience. Being abstract, these concepts are "verbally fixed," as Merleau-Ponty says (VI 95 / 67). Then they are put in absolute opposition to one another. The logical consequence is that we have a pure nothingness that is not, and a pure being that is. But since this pure nothingness is nothing, it collapses; it is in fact identical to being. As Merleau-Ponty says, "as absolutely opposed, being and nothingness are indiscernible" (VI 94 / 66). For Merleau-Ponty, Sartre's dialectic is only so called; it is in fact a philosophy of identity. Therefore, Merleau-Ponty's mixturism is opposed to Sartre's philosophy of identity, Sartre's, we might say, "ontological monism."[9] So we can see already that Merleau-Ponty's mixturism will have to be something like a philosophy of difference.

In order to understand *positively* the difference in which Merleau-Ponty's mixture consists, we can make use of three conceptual schemes from Merleau-Ponty's writings prior to "Eye and Mind." The *first* comes from Merleau-Ponty's 1942 *The Structure of Behavior*.[10]

In this book, Merleau-Ponty appropriates the idea of Gestalt—the form or the shape—in order to overcome the dualism of the physical and the psychological; here, too, even earlier than the lectures on the union of the body and soul, Merleau-Ponty speaks of a mixture (SC 212 / 197).[11] A mixture is, for Merleau-Ponty, a form, a relation of figure and ground (*fond*), a whole (SC 101 / 91). Here is the definition Merleau-Ponty provides of a whole: a whole is an indecomposable unity of internal, reciprocal determinations, meaning that if one of the parts changes, then the whole changes, and if all the parts change but still maintain the same relations among themselves, then the whole does not change (SC 50 / 47). In other words, not being the sum of its parts, the whole is not an aggregate; there are no *partes extra partes*, no parts outside of one another, and therefore the whole, the relation of figure and ground, is always ambiguous (cf. SC 138 / 127).

The *second* conceptual scheme for understanding this ambiguous or mixed relation of parts and whole comes from the beginning of his 1952 "Indirect Language and the Voices of Silence." In this text, Merleau-Ponty introduces Saussure's linguistics into French philosophy. Thanks to Saussure, we know that linguistic signs such as phonemes reciprocally determine one another by means of "diacritical differences." The reciprocal determination, which refers us back to the Gestalt, implies that Saussure cannot base language on a system of positive ideas. Due to the fact that Saussure is rejecting any sense of signs other than the diacritical, he must, according to Merleau-Ponty, be rejecting two ways of conceiving the whole and therefore two ways of conceiving the parts in relation to the whole. On the one hand, Merleau-Ponty tells us that the whole of language cannot be "the explicit and articulated whole of the complete language as it is recorded in grammars and dictionaries" (S 50 / 39). On the other, the whole of a language cannot be "a logical totality like that of a philosophical system, all of whose elements can be (in principle) deduced from a single idea" (S 50 / 39). Instead, as Merleau-Ponty says, "The unity [Saussure] is talking about is a *unity of coexistence, like that of the sections of an arch which shoulder one another*. In a whole of this kind, the learned parts of language have an immediate value as a whole" (S 50 / 39, my emphasis). Merleau-Ponty's comparison of the part-whole relation to that of the sections (*les éléments*) of an arch (*une voûte*) is illuminating. Clearly, if you change one stone, the arch falls; or, if you change all the stones but maintain the relations between them, then you still have the arch. The arch is not a mere aggregate of stones. Because the stones "shoulder" (*s'épaulent*) each other, each stone "has an immediate value as a whole"; each stone, in other words, is a "total part" (cf. OE 17 / 124). But this comparison implies that each stone, or, more precisely, each part, being a total part, is different from the whole and yet is identical to it. This sameness of identity and difference defines Merleau-Ponty's mixturism; indeed, in "Descartes's Ontology and Contemporary Ontology," Merleau-Ponty says that "the visible opens upon an invisible which is its relief or its structure and where the identity is, rather, nondifference" (NdC 1959–61, 195). To anticipate, we should note that the sameness of identity and difference is precisely how Foucault defines the modern reflection on finitude: "toward a certain thought of the Same—where Difference is the same thing as Identity" (*vers une certaine pensée du Même—où la Différence est la même chose que l'Identité*)" (MC 326 / 315, Foucault's capitalization).

In light of this definition of the modern reflection on finitude, it is not surprising that the *third* conceptual scheme for Merleau-Ponty's mixturism comes from his 1956 "Everywhere and Nowhere" (S 188 / 149). Here, Merleau-Ponty calls today's science "small rationalism" (*le petit rationalisme*), and any consideration of Merleau-Ponty's view of science must start here. Modern science, or small rationalism, takes its operations to be *absolute* (S 185 / 147). Today's science has become absolute by working on indices, models, and variables that it has made for itself. By contrast, what Merleau-Ponty calls "large rationalism" (*le grand rationalisme*), which is the philosophy of the seventeenth century, in a word, Cartesianism, takes its science and its artifices or techniques to be *relative*, relative to something *larger*, to God, to the "infinite infinite," or to the "positive infinite." Merleau-Ponty calls the positive infinite "the secret of large rationalism." The positive infinite is not numerical indefiniteness; rather, the positive infinite contains everything within itself: "every partial being directly or indirectly presupposes [the positive infinite] and is in return really or eminently contained in it" (S 187 / 149).[12] Every part's being eminently contained in God means that all beings resemble God. Or, there is a relation of analogy between the creatures and the creator. Resembling God, every partial being must be a total part. With large rationalism, we are very close to Merleau-Ponty's own thought,[13] and we have already noted that the concept of the mixture comes from Descartes.

Indeed, in "Everywhere and Nowhere" Merleau-Ponty expresses some nostalgia for large rationalism, telling us that large rationalism is "close to us." But most importantly, he says that large rationalism is the "intermediary through which we must go in order to get to the philosophy that rejects large rationalism." I do not think it is an exaggeration to say that "Eye and Mind" is Merleau-Ponty's precise attempt to go through this necessary intermediary of large rationalism to the philosophy that is opposed to it.[14] In "Eye and Mind," Merleau-Ponty is trying make today's science and its thought, which he calls "operationalism," *relative* once more to something other and larger than itself. In other words, he is trying to make us understand that "small rationalism" (which, again, is modern science) belongs to a "heritage" (S 186 / 148); small rationalism is a "fossil" of the "living ontology" found in large rationalism. But we cannot return to large rationalism; instead, its living ontology has to be "translated." In "Everywhere and Nowhere," Merleau-Ponty says: "Descartes said that God is conceived of but not understood by us, and this 'not' expressed a privation and a defect in us. The modern Cartesian translates: the infinite is as much *absence* as *presence*, which makes the negative and the human enter into the definition of God" (S 189 / 150, Merleau-Ponty's emphasis).[15] In a word, the translation makes the finite enter into God. Then the living ontology of large rationalism becomes the ontology of *sentir*, the ontology of sensibility that we see laid out in "Eye and Mind." So let us now turn to "Eye and Mind," in particular to part 3, which discusses Descartes's *Optics*.[16]

Descartes's Classical Ontology

According to Merleau-Ponty, in the *Optics* Descartes wants to conceive vision as thought, and *at the same time* to conceive vision as touch (OE 37 / 131). Thought and touch are not just two models of vision for Descartes, as some Merleau-Ponty commentators have claimed.[17] Vision in Descartes, according to Merleau-Ponty, is *a relation between* touch and thought. We can see the systematic relation between thought and touch in the following passage. This is Merleau-Ponty speaking: "Painting for [Descartes] is . . . a mode or a variant of *thinking*, where thinking is canonically defined as intellectual *possession* and self-evidence" (OE 42 / 132, my emphasis). Intellectual possession relates the immanence of consciousness, the cogito, or even the concept—and Merleau-Ponty always uses the word *immanence* to refer to the cogito—to grasping with the hand (NdC 1959–61, 180nA, 190).[18] For Merleau-Ponty, Descartes's conception of vision—or, more generally, *sentir*—as a relation between immanence and grasping involves two complementary mistakes (VI 168 / 127). These complementary mistakes are "fusion and survey" (VI 169 / 127).[19] If one conceives sensibility as fusion—as immediate grasping with the hand—one coincides with and touches pure facts; in this case, *sentir* takes place in an absolute proximity somewhere. If one conceives sensibility as survey (*survol*)—the view from nowhere—one intuits and sees pure essences; in this case, *sentir* takes place at an infinite distance everywhere (VI 169 / 127). In other words, according to Merleau-Ponty, Cartesian vision is at once too close to the thing seen and too far away from it. The mistakes reside *both* in the purity of touch, fusion and absolute proximity, *and* in the purity of vision (which in *The Visible and the Invisible* Merleau-Ponty calls the "kosmotheoros"; VI 32 / 15), survey and infinite distance.

This double mistake orients Merleau-Ponty's analysis of Descartes's conception of vision in the *Optics*. What Merleau-Ponty is trying to show is that Descartes's conception moves from one mistake to the other. And Descartes is able to make this move because he conceives of light as a mechanical cause. Descartes, according to Merleau-Ponty, considers not the light that we see but the light that makes contact with, the light that touches and enters into, our eyes from the outside (OE 37 / 131). In other words, Descartes considers light as an outside cause that makes real effects inside of us. Merleau-Ponty says, "In the world there is the thing itself, and outside this thing itself there is that other thing which is only reflected light rays and which happens to have an ordered correspondence with the real thing; there are two individuals, then, *connected by causality from the outside*" (OE 38 / 131, my emphasis). For Merleau-Ponty, the proximity of cause has two inter-related consequences.

First, and this is most important, causal contact eliminates resemblance; even the resemblance of the mirror image becomes a projection of the mind onto things. For the Cartesian, according to Merleau-Ponty, the image in the mirror is an effect of the mechanics of things. For Merleau-Ponty, because Descartes wants to conceive light on the basis of causality, a conception that requires no resemblance between a cause and an effect, we do not in fact have an image in vision, but rather a representation. A representation, such as an etching, works as signs do; signs in no way resemble the things they signify. Here, in the signs that do not resemble, we see the origin of the indices with which, according to Merleau-Ponty, today's science works (OE 9 / 121). Merleau-Ponty says, "The magic of intentional species—the old idea of efficacious resemblance so strongly suggested to us by mirrors and paintings—loses its final argument if the entire power of the picture is that of a text to be read, a text totally free of promiscuity between the seeing and the visible" (OE 40 / 132).[20] This citation brings us to the *second* consequence of Descartes's conception of light as causal contact: vision in Descartes is the decipherment of signs. This move to vision as decipherment, which starts with the conception of light through causality, leads to surveying thought (*la pensée en survol*). Since vision is the decipherment of signs, it *thinks* in terms of a flat surface; signs on a page, for instance (such as writing), are flat. But also, according to Merleau-Ponty, the representation, which is the effect of the mechanical light, immobilizes the figure so that it can be abstracted from the background. In the course from 1960–61, "Cartesian Ontology and Contemporary Ontology," Merleau-Ponty says, "This presence of the figure is all that [Descartes] retains from vision. The rest of the field is composed of such figures that are not present. The visible world is for me [that is, for a Cartesian] a world in itself upon which the light of the gaze is projected and from which the gaze cuts out [*découpe*] present figures. That eliminates the relation to the background, which is a different kind of relation." (NdC 1959–61, 229). This "different kind of relation," for Merleau-Ponty, would seem to be one of resemblance. In any case, Descartes takes only the external envelope of things, and this abstraction of the figure from the field is why for Descartes, according to Merleau-Ponty, drawing is what defines pictures (OE 42 / 132). Because the flat representation presents only the outlined figure, for Descartes, depth is a false mystery (OE 45 / 133). Cartesian space is in itself, one thing outside of another, *partes extra partes*, and thus depth is really width. If we think we see depth, this is because we have bodies (which are the source of deceptions); therefore depth is nothing. Or, if there is depth, it is my participation in God; the being of space is beyond every particular point of view (OE 46 / 134). God, then, who is everywhere and has no perspective, sees all things, without one hiding another; thus God creates, or better, draws, a "geometral," a surveying plan.[21] So, we can see now that Merleau-Ponty's analysis of vision in Descartes's *Optics* goes from fusion, at one extreme, to the other extreme, surveying thought (OE 48 / 134), the "kosmotheoros."

Merleau-Ponty's analysis is complicated, so I will reduce it to its most basic steps. According to Merleau-Ponty, Descartes starts from the conception of light as a cause contacting the eyes. The contact of light with the eyes is the absolute proximity of fusion. Because the contact with the eyes is causal, there is no resemblance between the image and the thing. Instead of images that resemble, we have signs. Signs are the figure without the background, immobile, and they are flat, like writing or a drawing. Vision, then, in Descartes becomes the decipherment of signs. And the decipherment of signs leads to the intellectual surveying plan, the geometral. The geometral is a drawing according to rectilinear perspective, with nothing hidden. It is surveying thought. Now, before moving on to what Merleau-Ponty says about painting, we should note two things about this analysis. *First*, the movement from fusion to survey is Merleau-Ponty's interpretation of Descartes's dualism of substances. Thus, how Descartes conceives vision in the *Optics* really concerns how the two substances (of course, mind and body) relate to one another. As the citation from "Cartesian Ontology and Contemporary Ontology" indicates, the two substances, according to Merleau-Ponty, interact by a *découper*, a cutting out or apart, a dividing. Therefore we can now provide more conceptual determination of Merleau-Ponty's mixturism. Like Sartre's ontological monism, Descartes's dualism of the division is opposed to Merleau-Ponty's mixturism. By contrast, for Merleau-Ponty, in sensibility there is an *indivision* between the sensing or activity and the sensed or passivity (OE 20 / 125). The move from division to indivision is Merleau-Ponty's translation, as mentioned earlier in "Everywhere and Nowhere" (S 189 / 150), of Descartes's ontology of substances into the ontology of sensibility. The *second* thing we must note, before we depart from Merleau-Ponty's analysis of vision in Descartes's *Optics*, is that Merleau-Ponty is making a distinction between image and representation. As we have seen, according to Merleau-Ponty, the positive infinite contains the properties of all partial beings in an eminent way; in other words, God possesses the same properties as creatures, only more so.[22] Thus, following the translation of the positive infinite, an image is always based on resemblance, on the sameness not of God and man, but of seeing and seen. By contrast, a representation is a sign; it involves no resemblance between the representation and the represented. So, we must anticipate, once again, the intersection with Foucault. In *Words and Things*, the final passage of the description of the structure of Velázquez's painting is: "This very subject [*ce sujet même*] — which is the same [*qui est le même*] — has been elided. And representation, freed finally from the relation [that of the same] that was structuring it [*l'enchaînait*], can give itself off as *pure* representation" (MC 31 / 16, my emphasis).

Merleau-Ponty's translation of large rationalism is not yet complete. Descartes, according to Merleau-Ponty, could not eliminate "the enigma of vision" (OE 51 / 135). Instead, the enigma is shifted from surveying thought, the thought of vision, to "vision in act" (OE 55 / 136). In other words, it is shifted to factual vision, to embodied vision. According to Merleau-Ponty, however, factual vision does not overthrow Descartes's philosophy. For Descartes, there is a limit to metaphysics. Since vision is thought united with a body, I can live it but not conceive it. As Merleau-Ponty says, "The truth is that it is absurd to submit the mixture [*le mélange*, of course] of the understanding and the body to the pure understanding" (OE 55 / 137). For Descartes, by being positioned (by being finite, in other words), we are disqualified from looking into both God's being and the corporeal space of the soul. Repeating a formula from "Everywhere and Nowhere," Merleau-Ponty in "Eye and Mind" calls this limit to metaphysics "the secret of the Cartesian equilibrium" (OE 56 / 137). Of course, just as we cannot return to large rationalism, this secret has been lost forever. Yet, as Merleau-Ponty stresses, since we are the composite of body and soul, there *must be* a thought of that composite. The thought of the composite would be as much opposed to small rationalism (operationalism or today's science) as to large rationalism (Cartesianism). As in the course from 1960–61, we can enter into this fundamental thought, into this philosophy "still to be made" only through art, only through the painter's vision (OE 61 / 138–39).

Thinking in Painting

The painter's vision, for Merleau-Ponty, goes beyond "profane" (OE 27 / 127) or "ordinary vision" (OE 70 / 142) to "the enigma of vision" (OE 64–65 / 140). Like Descartes's conception of vision, profane vision, according to part 4 of "Eye and Mind" (which is probably the most famous one), consists in two extreme views. On the one hand, there is the view from the airplane, which allows us to see an interval, without any mystery, between the trees nearby and those far away. On the other hand, there is "the sleight of hand" by means of which one thing is replaced by another, as in a perspective drawing (OE 64 / 140). With these two views, once again, we have the proximity of fusion (the contact through the hand) and the infinite distance of surveying thought (the distance from the airplane). The phrase "sleight of hand" translates Merleau-Ponty's *escamotage*, which means to make something disappear by a skillful maneuver;

"maneuver" literally means using the hand, which is why I rendered *escamotage* as "sleight of hand." But *escamotage* is also etymologically connected to the French word *effilocher*, which means to unravel or untie something that has been woven together. We can now see that both the sleight of hand and the view from the airplane separate things and make them be *partes extra partes*. This maneuver and view are the opposite of the interweaving in which the enigma of vision consists.

Here is Merleau-Ponty's definition of the enigma of vision: "The enigma is that I see things, each in its place, precisely because they eclipse one another; it is that they are rivals before my sight precisely because each one is in its own place. The enigma is their known exteriority in their envelopment, and their mutual dependence in their autonomy. Once depth is understood in this way, we can no longer call it a third dimension" (OE 64–65 / 139). We can see the oxymoronic formulas by means of which Merleau-Ponty is defining the enigma: exterior—known, they are *partes extra partes*—and yet enveloped, dependent in autonomy. But we can also see the reversibility. Each thing is in its own place, exterior to one another, because they hide one another in envelopment; they are rivals, mutually dependent, because each is in its own place, autonomous. While for Descartes depth was a false problem, for Merleau-Ponty, as this quote indicates, depth is the whole question. For him, depth is the first dimension or the source of all dimensions, "dimensionality" (OE 48 / 134), "voluminosity" (OE 27 / 127), the "there," the "one same space" (OE 85 / 147), the "one same being" (OE 17 / 124); depth is the experience of the reversibility of dimensions, of a global "locality," where all the dimensions are at once (OE 65 / 140). Now, his discussion of depth (*la profondeur*) in "Eye and Mind" refers us back to his early work in *The Structure of Behavior* on the Gestalt, the relation of figure and ground (*le fond*). Therefore, we can see how Merleau-Ponty is proceeding here (in part 4). With the enigma of vision we have depth, and therefore we have the background. Now we need the figure.

For Merleau-Ponty, the figure is generated by color and line. But color and line, like all the other dimensions, are not based on a "recipe," as Merleau-Ponty says, for the visible. It is not a question of adding other dimensions to the two of the canvas. The lack of a recipe means that for Merleau-Ponty painting or, more generally, pictures do not imitate nature. Merleau-Ponty is rejecting the traditional concept of imitation, which implies an external relation between the painter and something outside of him- or herself. For Merleau-Ponty, the painter is not viewing something else from the outside. Instead, the painter *is born in the things* by the concentration and coming to itself of the visible. This "being born in [*dans*] the things" is what Merleau-Ponty means when he speaks of the picture being "auto-figurative" (OE 69 / 141). But what is most important about this discussion in part 4 of "Eye and Mind"—it seems to me that pages 69–72 / 141–42 are the heart of the essay; they overlap with the final pages of chapter 4 of *The Visible and the Invisible*[23]—is that not only is the painter born in the things, but so is the writer, or better, the poet. Here, through the idea of auto-figuration, Merleau-Ponty is trying to bring the language arts back to painting, back to the visible.[24] *First*, Merleau-Ponty refers to Apollinaire, who said that there are phrases in a poem that do not appear to have been created but that seem "to have formed themselves" (OE 69 / 141). *Then, second*, Merleau-Ponty quotes Henri Michaux, saying that Klee's colors seem to have been born slowly upon the canvas, to have emanated from "a primordial ground" (*un fond primordial*), "exhaled at the right spot like a patina or mold." *Between* these two comments we have an *et*, an *and*, which implies a comparison, or better, a *compatibility* between the colors forming themselves on the canvas and the words forming themselves on the page, a compatibility between the eye that sees and the eye that reads. Here we must also refer to the intersection with Foucault. On the one hand, Apollinaire of course composed his poems as calligrams, the calligram being what Magritte "unmakes" according to Foucault in *This Is Not a Pipe*. On the other, in chapter 9 of *Words and Things*, Foucault says that an *et* connects the doubles that define man's ambiguous existence. The *et* means that *Merleau-Ponty* wants the painter and the poet—in a word, man—not on the inside *of God* (this would be large rationalism), *but on the inside of the visible*. Merleau-Ponty's definition of art shows us that this *et* implies a mixture, an ambiguous relation of light and feeling, of the visible and the invisible. Art, for Merleau-Ponty, is not a "skillful relation, from the outside, to a space and a world." Instead, "art is the inarticulate cry, the voice of light," "la voix de la lumière" (OE 70 / 142). In the course from 1961, "Cartesian Ontology and Contemporary Ontology," Merleau-Ponty reproduces Valéry's poem "Pythie," which speaks of a voice of no one, the voice of the waves and the woods, which is literature and the unveiling of the visible, the speech of things. Merleau-Ponty comments on this poem by saying that "the visible and what the poem means [are] interwoven [*entrelacés*] (NdC 1959–61, 186).[25]

In "Eye and Mind," Merleau-Ponty provides a remarkable example of this interweaving in a painter's vision (not the profane vision) of a swimming pool.[26] It is clear that, in this description of the view of a swimming pool, Merleau-Ponty is still concerned with a figure-ground relation, since he is speaking about the bottom (*le fond*) of the pool. Here is the description:

> If I saw, without this flesh, the *geometry* of the tile, then I would stop seeing the tiled bottom as it is, where it is, namely: farther away than any identical place. I cannot say that the water itself—the aqueous power, the syrupy and shimmering *element*—is *in* space; all this is not somewhere else either, but it is not in the pool. It dwells in it, is materialized there, yet it is not contained there; and if I lift my eyes toward the screen of cypresses where the web of reflections plays, I must recognize that the water visits it as well, or at least sends out to it its active and living essence. This inner animation, this radiation of the visible, is what the painter seeks beneath the names of depth, space, and color. (OE 70–71 / 142, my emphasis)

Merleau-Ponty selects the vision of a swimming pool because, it seems, any swimming pool has to have depth so that one might be able to swim in it. The depth is the water, which is not in space or in the pool; the water "dwells there," as Merleau-Ponty says, but "dwelling" (*habiter*) means that the water is not contained in the pool but is itself the container. Or, as Merleau-Ponty says here, it is an "element." Now, in *The Visible and the Invisible*, Merleau-Ponty also calls the flesh an element, saying, "to designate the flesh, we would need the old term 'element,' in the sense it was used to speak of water, air, earth, and fire, that is, in the sense of a general thing, midway [*mi-chemin*] between the spatio-temporal individual and the idea" (VI 184 / 139). Without the flesh of the water, we would be able to grasp the tiles with our hands and hold them in one identical place, but then we would not see their geometry, or more precisely, geometry per se. The flesh allows us to see geometry, since the water's distortions function as a sort of variation of the spatio-temporal individual. The variation makes the geometry be "farther away than any identical place." But, being midway, the water makes the geometry be not so far away as to exist in a second world of forms without any support from the visible (cf. OE 91 / 149); again, we see here that Merleau-Ponty's thought is an anti-Platonism. The geometry reaches only as low as the bottom of the syrupy element and only as high as the screen of cypresses.

You can see, I hope, that with this description of the swimming pool Merleau-Ponty is no longer speaking of voice. The geometry of the tiles refers us to line. It is well known that Merleau-Ponty says that modern painting contests the "prosaic line," the line between a field and a meadow, which the pencil or brush would only have to reproduce. Again, we can see that Merleau-Ponty is not interested in the traditional idea of art as imitation or reproduction. It is also well known that in this context Merleau-Ponty turns to Klee again. For Klee, according to Merleau-Ponty, the line is the genesis of the visible, and then, still according to Merleau-Ponty, Klee "leaves it up to the *title* to designate by its prosaic name the being thus constituted, in order to leave the painting free to function more purely as a painting" (OE 75 / 143, Merleau-Ponty's emphasis). In the course "Cartesian Ontology and Contemporary Ontology," Merleau-Ponty also speaks of the role of the title in Klee, saying that the title "disburdens the picture of resemblance [here Merleau-Ponty means imitation] in order to allow it to express, to present an alogical essence of the world which . . . is not empirically in the world and yet leads the world back to its pure ontological accent, it puts in relief its way of *Welten* [worlding], of being world" (NdC 1959–61, 53). This citation means that the title designates *the thing* whose genesis the painting is showing us—without the painting imitating that thing. So, Merleau-Ponty says in "Eye and Mind" that Klee has painted two holly leaves *exactly* in the way they are generated in the visible, in the way they "holly leave," we might say, and yet they are indecipherable precisely because the painting does not imitate the empirical object called holly leaves; the title instead designates this empirical object which has been generated. It is important that Merleau-Ponty does not say that the title in Klee *denies* that the painting is of holly leaves. Klee does not say, "This is not two holly leaves," "ceci n'est pas deux feuilles de houx." The title *affirms* that they are indeed holly leaves, which implies that the title, like the phrases in the poem, like the geometry of the tiles at the bottom of the pool, is the outgrowth of the genesis, its final stage, its patina or mold, its exhalation. We might go so far as to say that the relation between the title and the painting in Merleau-Ponty is that of a calligram: the lines emerge from depth and then they become words that still *resemble* the depth from which they came. Thus, recognizing the weaving of words into things, we can interweave the two quotations Merleau-Ponty uses to frame part 4 of "Eye and Mind." The first, which completes part 4, is from Klee: "I cannot be grasped in [*dans*] immanence," in the immanence, that is, of consciousness, of the cogito, of thought (OE 87 / 148). The second quote, which completes part 3, of course comes from Cézanne: the painter "thinks in [*en*] painting" (OE 60 / 139).[27]

Conclusion: Man and His Doubles

The preposition in this phrase from Cézanne, "pense *en* peinture," expresses, for Merleau-Ponty, the indivision of the invisible and the visible, of words and things. What is at issue in this philosophy that comes from painting is the connection between these two (OE 64 / 140), the "between," the *entre-lacs*, the inter-weaving, as Merleau-Ponty says in *The Visible and the Invisible*. Being a "thought of the inside,"[28] Merleau-Ponty's philosophy is always trying to move *into* this "between." This interiority is why Merleau-Ponty rejects the traditional concept of imitation, in which the imitation is between two things outside of one another. Yet, despite the criticism of imitation, we must say that, while depth (*la profondeur*) is no-thing, there is a resemblance between the figure and the ground (*le fond*). If we are correct about the conceptual schemes for Merleau-Ponty's mixturism, then we must recognize that the logic of the positive infinite implies a relation of eminence between the figure and the ground. Of course, again, what Merleau-Ponty is speaking about is *not* traditional imitation, not a copying relation, but resemblance and images. In "Eye and Mind," Merelau-Ponty's thoughts about resemblance are especially guided by the specular image (OE 28 / 128). Resemblance therefore, for Merleau-Ponty, seems to work in this way. In a mirror, I see my flesh outside, and as outside, I recognize my inside (an inside that, if I am a child, had hitherto been confusedly felt affects). But this recognition does not occur before the mirror image, and it occurs only on the basis of that specular image, which is outside. Then I can transfer this recognized inside to other outsides, which are like the specular image I saw of myself outside. In other words, on the basis of the specular image, I can attribute my inside to another's flesh, even though the inside of another's flesh remains invisible, even though it is *Nichturpräsentierbarkeit* (VI 292 / 238–39).[29] As Merleau-Ponty says, "They [that is, the image, the picture, and the drawing] are the inside of the outside and the outside of the inside, which the duplicity [*duplicité*] of sensibility makes possible and without which we would never understand the quasi-presence and imminent visibility which make up the whole problem of the imaginary" (OE 23 / 126). It is significant, of course, that here Merleau-Ponty is alluding to Lacan's mirror stage, about which he had lectured in 1949, *and* that he speaks of the imaginary and not of the symbolic.[30] But what we must stress is that, for Merleau-Ponty, the vision of the painter "gives visible existence to what profane vision believes to be invisible. . . . This voracious vision, reaching beyond [*par delà*] the 'visual givens,' opens upon a texture of Being of which the discrete sensorial messages are only the punctuation or the caesura" (OE 27 / 127). Because painting reaches beyond and gives visible existence to what was invisible, for Merleau-Ponty there is only ever "the invisible of the visible" (VI 300 / 247). The invisible is always relative to the visible and is always on the verge, imminently, of being visible, of coinciding with the visible (cf. VI 163 / 122–23). The invisible is never a teeming presence but always on the horizon of the visible (VI 195 / 148). And even if we can speak of a "blind spot" (VI 300–301 / 247–48), an "impotence" (*impuissance*) of vision (VI 194 / 148), Merleau-Ponty always conceives it, not on the basis of noncoincidence, but on the basis of coincidence, not on the basis of blindness, but on the basis of vision, not on the basis of impotence, but on the basis of the "I can."[31] Here, in the question of power, we have the subtle shift in emphasis between Merleau-Ponty and Foucault. The subtle shift in emphasis means that all the prepositions in Merleau-Ponty, the "to" (*à*), the "in" (*en*), the "within" (*dans*), the "beyond" (*par-delà*), and the "between" (*entre*), in short, the inside, have the signification of resemblance. If we are going to have a strict difference between immanence and transcendence, then the resemblance relation implies that Merleau-Ponty is *not* a philosopher of immanence, but a philosopher of transcendence. We should recall again what Klee says: "I cannot be grasped in immanence."

What, or better, who is the emblem of transcendence in Merleau-Ponty? Who is the "between"? Between the two extremes of the distant view from the airplane and the up-close grasp of the sleight of hand, between survey and fusion, between the screen of cypress tress and the bottom of the pool, there is the vision of the eyes. The eyes see that things are not flat and juxtaposed; one thing stands behind another and is therefore obscure and hidden. But the eyes see in this way only if the body is upright, with feet on the ground.[32] The verticality of the upright body is not, of course, vision absorbed into the cogito, as in Descartes. Nevertheless, I think that it is necessary to recognize that whenever Merleau-Ponty speaks of verticality, as he does so often in the working notes to *The Visible and the Invisible*, he is privileging the human body and its uprightness (cf. VI 325 / 271–72). In "Eye and Mind," he says, "This interiority [that is, the indivision of the sensible and the sensing] does not precede the material arrangement of the *human body*, and it no more results from it" (OE 20 / 125, my emphasis). For Merleau-Ponty, the "fundamental of painting, perhaps of all culture" (OE 15 / 123) is the human—and not the animal—body. The upright human body is the "between" of survey and fusion, the *mi-lieu*, the *mi-chemin* between essence and fact (cf. VI 328 / 274). Since the human body is visible, the human is the figure standing out from the ground of the visible; as the figure, man can be studied as an empirical positivity.[33] *And*, since the human body sees, the human resembles the ground of the visible; as the ground, man can as well be taken to be the transcendental foundation. As Merleau-Ponty says, "the manifest visibility [of things] doubles itself [*se double*] in *my* body" (OE 22 / 125, my emphasis). Therefore we must conclude by saying that Merleau-Ponty's thought, his "mixturism," is defined by the phrase "l'homme et ses doubles."

7

Noli me tangere
A Fragment on Vision in Merleau-Ponty

It is obvious that, in *The Birth of the Clinic*, Foucault's use of the phrase "visible and invisible" alludes to Merleau-Ponty. If someone knows anything about Merleau-Ponty, that person knows his description of the touching-touched relation. Yet it seems to me that one must always recall that Merleau-Ponty's final published work is "Eye and Mind," not "Hand and Mind." Merleau-Ponty—and here again Foucault is quite close to Merleau-Ponty—is a philosopher of vision. To demonstrate this point, let me bring forward two short quotes from Merleau-Ponty's unfinished *The Visible and the Invisible*. First, in chapter 2, Merleau-Ponty says, "To be sure, our world is principally and essentially visual; one would not make a world out of scents or sounds" (VI 115 / 83); then, in chapter 4, "vision comes to complete the aesthesiological body" (VI 202 / 154). But perhaps the strongest evidence for this claim lies in the fact that Merleau-Ponty, in "Eye and Mind," criticizes the Cartesian theory of vision because it is modeled not on seeing but on touch, not on action at a distance but on action by contact (OE 37). Let me be more precise. What Merleau-Ponty is attempting to do, perhaps in all of his work, is to reconceive sensing, to use Merleau-Ponty's French, to reconceive *sentir*. But to do this, it seems to me, one must privilege a particular sense, more precisely, a specific experience. I am arguing that Merleau-Ponty privileges vision—and one should not overlook the ambiguity of this term.

The privilege that Merleau-Ponty gives to vision does not mean that "the prejudice of presence" dominates his thinking.[1] Merleau-Ponty is opposed to this prejudice because he sees that it consists in fact in a prejudice of purity. This prejudice of purity can function in two ways for Merleau-Ponty. One can conceive sensing, that is, to repeat Merleau-Ponty's French, *sentir*, as fusion, in which case one touches pure facts—here, *sentir* takes place in an absolute proximity somewhere. Or, one can conceive *sentir* as a survey (*survoler*), in which case one sees pure essences—here, *sentir* takes place at an infinite distance everywhere (ubiquity) (VI 168–69 / 127). The prejudice resides in both the purity of touch, fusion and absolute proximity, and the purity of vision (which Merleau-Ponty also calls the "kosmostheoros"), survey and infinite distance. These two sides are complements, and Merleau-Ponty gives us an oxymoronic expression to help us overcome them: "palpation with the eyes."[2]

Although Merleau-Ponty does not say what I am about to say, it seems to me that it is implied by his rejection of *sentir* conceived as fusion or coincidence. The first thing one must notice about the experience of vision is that it involves a prohibition: I must not touch the thing seen, or I must keep it at a distance. Vision is based on this prohibition because if I apply the thing I am looking at right on my eyeball, I cannot see it. Therefore, to repeat my title, "Noli me tangere!" Thanks to this prohibition, then, "we see things themselves," which is the first sentence of *The Visible and the Invisible*. But we can still ask: Why or how is it possible to see things *themselves*? Due to the prohibition of contact, I am not fusion. Therefore, if I am, as well, not a *kosmostheoros*—a view from nowhere—if my eyes see (and not my mind from an infinite distance), then my eyes are visible things too; they can be seen. Vision involves, therefore, a fundamental passivity (VI 183 / 139): the seer can be seen. Since I am visible, I am related to or even resemble the things I see, which allows me to see them *themselves*. Yet again, if my eyes see, if one is standing upright, vertical, with feet on the ground, things are not flat and juxtaposed.[3] One thing stands behind another and is therefore obscured and hidden. Here, Merleau-Ponty appeals to Husserl's idea of horizon (VI 195 / 148). Like a horizon, which recedes as one approaches, which remains invisible, which cannot be touched, the distance in vision is "for good" (cf. VI 122 / 89). Merleau-Ponty's idea of a distance that is "for good" probably is the source of Foucault's work on vision. Perhaps with this distance Merleau-Ponty too was on the verge of conceiving singularities that could not be recognized. We must leave this question to the side.

There is a more obvious question. What about touch in Merleau-Ponty? Merleau-Ponty indeed speaks about contact, touch, and the touching-touched. In the famous fourth chapter to *The Visible and the Invisible*, Merleau-Ponty discusses touch twice. The first time (VI 175–76 / 133) occurs when Merleau-Ponty is trying to explain how it is that the gaze seems to possess already, as if in a "pre-established harmony," visible things. More than the eyes, the hands are seen and touched. The passivity that allows for an in principle "kinship" between sensing and sensed can be understood more easily, therefore, through touch, and Merleau-Ponty says that the palpation of the eyes is a "remarkable variant" of this "closer" palpation by the hands (VI 175 / 133). The context of the second time in which Merleau-Ponty discusses touch is very specific. In discussing the experience of others, Merleau-Ponty has opened up the possibility of ideas and thought, or the mind. He then suddenly worries (if you ask me) that he has relapsed into the position of *Phenomenology of Perception*, placing a kind of tacit thought or cogito at the foundation of the body (VI 191 / 145). On the basis of certain famous working notes, we know that Merleau-Ponty was trying to avoid, in *The Visible and the Invisible*, anything like a tacit cogito. But then, after this worry, he feels obligated to re-examine his notion of the flesh. He says, "To begin with, we spoke summarily of a reversibility of seeing and the visible, of touching and the touched." He continues, "It is time to emphasize that it is a reversibility always *imminent* and never realized in fact. My left hand is always on the verge [*sur le point*] of touching my right hand touching the things, but *I never reach coincidence*" (VI 194 / 147, my emphasis). I know this will sound strange, but I think the only way to interpret this *sur le point* is to say *now* that the closer palpation of the hands is a "remarkable variant" of vision. The imminence implies that there is a distance there, between the two hands, which is "for good," and therefore the opening of a horizon and invisibility is there too. In fact, we can go farther with this interpretation. The touching-touched experience is what psychologists have called a double sensation. But this sensation can really be double, can be two and not one, only if there is no fusion. If there were fusion or coincidence, then *either* everything would be touching, fused into activity, *or* everything would be touched, fused into passivity (cf. VI 163 / 122), but not *both* touching-touched. Even in relation to this famous, tangible example, we would have to say that there is a prohibition against touch, even here we would have to say, "Noli me tangere!" Indeed, Merleau-Ponty says almost as much. He discusses touch the second time because he is concerned about the genesis of ideas. The imminence of coincidence implies that there is always hiddenness and invisibility in this experience, in the flesh, and this hiddenness is the origin of ideas. So, Merleau-Ponty says:

> [ideas] could not be given to us as ideas except in a carnal experience . . . they owe their authority, their fascinating, indestructible power, precisely to the fact that they are in transparency behind the sensible, or in its heart. Each time we want to get access to the idea immediately or *lay hands on it* [*mettre la main sur elle*], or circumscribe it, or see it unveiled, we really feel that the attempt is *misconceived* [*la tentative est un contre-sens*]. (VI 197 / 150, my emphasis)

Even if we must say that Merleau-Ponty's imminence of coincidence is not stubborn enough, we can now ask Foucault's question: "Who are we?" I think the answer to this question must remain open, in constant differentiation and alteration. If the developed forms of human experience are concerned primarily with the goal of a recognizable self-identity, then I think that these forms must be reduced, reduced *down* into singularities, which would be more immanent than the self, and reduced *up* into ideas, which would be more transcendent than the self. In other words, the reduction of the autonomy of the developed forms of human experience means a transformation of our human thinking into a nonhuman mode, into a nonhuman (and not human) experience. I think *this* reduction is the only way out of the form of life called "man," and therefore the only way into what calls for thinking: an experience of what is above or below man. The title of my comments, "noli me tangere," of course, is the Latin translation of the Greek "me mou aptou," which is found in John's Gospel (20:17).[4] The risen Jesus, that is, a Jesus who is still alive while dead, tells Mary Magdalene, "Don't touch me!"; she may look upon him only. It seems to me that only the prohibition, the impotence of touch and the potency to see, preserves the distance that makes him other than human. It seems to me that this distance, given in an exemplary way in vision, is the condition of all alterity. It turns us into the followers (*les suivants*).

8

Un écart infime (Part III)
The Blind Spot in Foucault

All of Merleau-Ponty's thought consists in a mixturism. The eye, vision, in Merleau-Ponty mixes together passivity and activity. Yet passivity, in Merleau-Ponty, seems to amount to a sort of blindness. Indeed, in two working notes to *The Visible and the Invisible* (from May 1960), Merleau-Ponty speaks of a *punctum caecum*, a "blind point" (VI 300–301 / 247–48). If we think quickly of Foucault's analysis of *Las Meninas*, with which *Words and Things* opens and where we are heading in this essay, we see that it, too, concerns a "blind point." Merleau-Ponty's thought, therefore, seems very close to that of Foucault, and, of course, it is. After all, Merleau-Ponty died in 1961, and two years later, in 1963, Foucault described his *Birth of the Clinic* as the re-examination of "the originary distribution of the visible and the invisible" (NC vii / xi). Yet, as we have seen in Chapter 6, there is a subtle shift in emphasis between Merleau-Ponty and Foucault. This subtle shift in emphasis is the starting point here.

For Merleau-Ponty in "Eye and Mind," the vision of the painter reaches beyond the visual givens and gives visible existence to what is invisible, which implies that invisibility is always *imminent* visibility (OE 23 / 126), the invisible at the horizon of the visible (VI 195 / 148). So, even if we can speak of a "blind spot," an "impotence" (*impuissance*) of vision (VI 194 / 148), Merleau-Ponty always conceives it not on the basis of noncoincidence but on the basis of coincidence, not on the basis of blindness but on the basis of vision, not on the basis of impotence but on the basis of the "I can," finally, not on the basis of something like an absolute invisibility but on the basis of "the non-mediated presence which is not something positive" (VI 302 / 248). Because for Merleau-Ponty invisibility is always relative to the visible, because coincidence is always partial, all the prepositions in Merleau-Ponty—"to" (à), "in" (en), "within" (∂ans), "beyond" (par-∂elà), and "between" (entre), in short, what he calls "the inside," have the signification of resemblance. If we are going to make a strict conceptual difference between immanence and transcendence, the resemblance relation implies that Merleau-Ponty is *not* a philosopher of immanence, but a philosopher of transcendence. But even more, the resemblance relation implies that the upright human body is the "between" of survey and fusion, the *mi-lieu*, the *mi-chemin* between essence and fact (cf. VI 328 / 274). Since the human body is visible, the human is the figure standing out from the ground of the visible; as the figure, man can be studied as an empirical positivity. *And*, since the human body sees, the human resembles the ground of the visible; as the ground, man can as well be taken to be the transcendental foundation. As Merleau-Ponty says, "the manifest visibility [of things] *∂oubles* itself [*se ∂ouble*] in *my* body" (OE 22 /125, my emphasis). Therefore, Merleau-Ponty's thought, his "mixturism," is defined by the *et* in "l'homme et ses doubles."

In Foucault, there is no resemblance relation between the doubles. *Un écart infime*, a minuscule hiatus, *un ∂épli*, an "unfold," divides the doubles and yet relates them in a relation of incompatibility. Incompatibility means that between words and things, between the visible and the invisible, there is a kind of battle, a battle between the visible and the invisible, but not a relation of resemblance. In light of the battle, what, then, does painting do in Foucault? Speaking of Blanchot's fiction in "The Thought from the Outside," Foucault says — and, as we shall see, Foucault will call the space of the painting "fictitious" (MC 26 / 10) — "fiction consists *not* in making us see the invisible, *but* in making us see *how much* the invisibility of the visible is invisible."[1] For Merleau-Ponty, painting makes us see the invisible; for Foucault, painting — or as he says here, fiction — makes us see how much the invisibility of the visible is invisible. For Foucault, the invisible is never an imminent visible on the horizon. Foucault's "blind point" is a kind of "a-perspectivism," in the literal sense;[2] there can be no in-spection of this spot; it cannot be turned into spectacle; and thus no change of perspective would allow us to see it. And yet, the invisible in Foucault is not absolutely absent;[3] it is diffracted into singular visibilities and then has "a teeming presence" like death (*une présence fourmillante*).[4] This teeming presence is what we are going to see in Foucault's famous (or infamous) analysis of Velázquez's painting. I am going to reconstruct the analysis—the title of this first chapter is "Les Suivantes" ("The Handmaidens," or "The Followers"; I will use the French title throughout because the title in the English translation, "Las Meninas," fails to capture nuances that will be key to me here)—in order to show how what I am calling the "blind spot," the impossibility of vision, is connected to life, to power, and thus to thinking.[5] It is indisputable, it seems to me, that in *Foucault* Deleuze has given us the most philosophically interesting reading of Foucault. Deleuze tells us that "one thing haunts Foucault, and that is the question of thinking, the question shot by Heidegger and taken up by Foucault, *the arrow par excellence.*"[6] In *This Is Not a Pipe*, Foucault suggests that painting itself is a form of thinking. As we will see, "thinking in painting" consists in diffraction. As we will also see, the image of the arrow will guide us through Foucault's analysis of the Velázquez painting. But why should we focus on this analysis? It seems to me, and this is our starting point, that the analysis of *Las Meninas* has an unusual status in Foucault's thinking overall.[7]

The Singular Status of "Les Suivantes"

What is immediately striking about "Les Suivantes" is the absence of proper names. Moreover, neither of the chapter's two sections bears a title. In the first section, the only proper name to appear (at the very end) is that of Velázquez's teacher, Pachero. Foucault does not even mention the name *Velázquez* until the beginning of the second section.[8] Here, at the beginning of the second section, he says—and these comments are perhaps what is best known about the chapter— "Perhaps it would be better to fix once and for all the identity of the characters either presented or indicated by the painting" (MC 24 / 9). As is well known from historical anecdote, the figures in the painting are King Philip IV and his wife Mariana, with their child, the Infanta Margarita. The king and the queen appear in the reflection in the mirror in the back of the painting, and the child is in the center of the painting, with her handmaidens, court jesters, and courtesans around her. These proper names—King Philip IV and Mariana, Margarita, and so forth—would be, as Foucault says, "useful landmarks" (*utiles repères*), and would help us to avoid "ambiguous designations." But, for Foucault, proper names are only an "artifice" in the "play" between names or words and the painting or the visible. He says, "They [that is, proper names] allow us to point the finger [*montrer du doigt*], that is, they make us pass surreptitiously from the space in which one speaks to the space in which one gazes [*on regarde*], that is, they allow us to close one conveniently over the other as if the two were adequate" (MC 25 / 9). Instead of this artifice of the proper name, Foucault advocates another artifice: "it is necessary to erase the proper names"; "it is necessary therefore to pretend [*feindre*] not to know who is being reflected at the bottom of the mirror and question this reflection just in terms of its existence" (MC 25 / 9). As far as I know, "Les Suivantes" is the only analysis of a painting in which Foucault adopts this pretence of not knowing the names of the figures in the painting he is analyzing. Indeed, we must pretend not even to know the name of the painter. In "Les Suivantes," Foucault mentions Velázquez's name only three times in the entire analysis; all three times occur in the second section. In contrast to what we see in "Les Suivantes," in *Manet's Painting*, a lecture that dates from roughly the same period, Foucault constantly uses Manet's name.[9] But why employ this pretence of ignorance, why this reduction of knowledge, why this other artifice against the artifice of the proper name?

In order to answer this question, let us examine the placement of this chapter in *Words and Things*. Foucault had previously published it in *Mercure de France* in 1965, a year before the publication of *Words and Things*.[10] He therefore had to decide where to place the text (which he also revised) in the book. He put it at the beginning of part 1, which primarily concerns the classical epoch (the epoch from Descartes to Kant)—the epoch, of course, of representation. Foucault locates the analysis here because Velázquez's painting decomposes (according to Foucault) all the elements of representation in order to make the painting into a representation of representation.[11] The reason for its placement becomes obvious: it opens part 1 because Velázquez's painting opens up the classical epoch.

Yet the analysis of the Velázquez painting returns in part 2, which concerns "the limits of representation," in other words, the transition from the classical epoch to the modern epoch, to the nineteenth and twentieth centuries. Part 2 especially introduces the idea that defines the modern epoch, that is, the idea of man. In chapter 9 of part 2, "Man and His Doubles"—a title that alludes to Artaud's work on theater—in a section called "The Place of the King," Foucault introduces what he calls "the character" (*le personage*) of man. He introduces this character as "a Deus ex machina" (*un coup de théatre artificiel*) into the "great classical play of representation." And yet, he says, "we would like to recognize the prior law of this play in the picture of the *Meninas*" (MC 318/307). We must stress the word *prior* (*préalable*) here. The analysis of the Velázquez painting makes visible, by means of the pretense of erasing proper names, the prior law that makes the play of representation possible. But, since Foucault is able to introduce man into the classical play of representation, the same law makes possible the modern play of the doubles of man.

Now let us return to the placement of "Les Suivantes" in part 1. In fact, it does not directly introduce the classical epoch; immediately after it falls the chapter called "The Prose of the World," whose title alludes to Merleau-Ponty but which concerns the Renaissance period. If we keep in mind the chapter's placement before the Renaissance epoch and not between the Renaissance and the classical, we must conclude that the analysis also makes visible the prior law that makes possible the medieval play of the same. This prior or a priori law, this essential structure, is the fundamental condition for all of these epochs and perhaps for the epochs that are still to come. "Les Suivantes" has an unusual status in Foucault's thinking precisely because, through the analysis of Velázquez's *Las Meninas*, this essential structure (which, we must say, is something more than a *historical* a priori) becomes manifest. But, following the opposition that Foucault makes in *The Archeology of Knowledge*, we see that what becomes visible here is also not a *formal* a priori.[12] By erasing the proper names, Foucault is trying to make visible the in-formal, the un-mappable or un-measurable space (*l'espace irreparable*; cf. MC 21/5) in the mirror relation of seeing and seen—we shall turn to the mirror relation in a moment.[13] In other words, Foucault is trying to make it be visible that "we live in the myriad of lost events *without originary landmarks* [*repères*] *or coordinates.*"[14] The artifice of erasing the proper names places us in the "non-place."[15] Now, of course the epochs that Foucault dissects in all his works are epochs of thinking (cf. MC 353/342). It is possible to claim, therefore, that, with the analysis of the Velasquez painting, we have before us the prior law of all thinking, the *dépli* in which it is possible once more to think the unmeasurable space that, being prior to all thinking, is itself impossible to think (cf. MC 7/xv). Now, let us turn to "Les Suivantes" in order to see how this law, which is the condition for the possibility of all thinking, while itself being impossible to think, functions.

An Overview of Foucault's Analysis of Velázquez's *Las Meninas*

As we have noted, Deleuze calls the question of thinking "the arrow *par excellence*." In fact, the image of the arrow appears in many of Foucault's writings.[16] In "Les Suivantes," in particular, it appears indirectly, through what Foucault calls the "sagittal" lines exiting from the Velázquez painting. The word *sagittal* literally means "arrowhead." By focusing, therefore, on Foucault's discussion of the sagittal lines, one can determine three major phases in his analysis of the painting, even though, as I said, the chapter is divided into two sections.[17] Following the sagittal lines means focusing not on movements back and forth across the painting but on movements that emerge from the back of the painting to the front.

The first major phase occurs in the first half of the first section. It concerns what Deleuze in his 1967 essay on structuralism—like his book, this essay provides a lot of insight into how one should read Foucault—calls "the empty square" (*la case vide*). The phrase refers to the square on a chess board, but we need to keep in mind, when we turn to Foucault's analysis, that the French word *cadre*, "frame," etymologically refers to the Latin for square, *quadrum*. In any case, according to Deleuze (who refers explicitly to Foucault's analysis of the Velázquez painting[18]) the empty square is an object, an agency, or a point that, while being immanent to both, allows two series, such as signifiers and signifieds or words and things, to communicate without being unified (QRS 259–60 / 185). The empty square is not synthetic but rather "symbolic." Deleuze appropriates this term from Lacan, but, as we shall see, Foucault himself uses the term *symbolic* in "Les Suivantes." In Foucault's analysis, the two series that are being symbolically related are the spectacle and the gaze, or the object of vision and the subject of vision. In the Velázquez painting, the principal figures in which the spectacle consists gaze at the spectator, turning him into a spectacle at the very moment he is gazing at them. As Foucault says, the painting is a *spectacle-en-regard*, a spectacle that is gazing (MC 29 / 14).[19] The Velázquez painting, in other words, sets up a mirroring relation between the spectators and the spectacle, which is then centered in the mirror at the back of the painting. The empty square is that mirror; in it, we can see the sovereigns, in particular, the king; so the empty square is "the place of the king" (cf. MC 318 / 307). Because the mirror is the empty square, we could say that

Foucault's entire analysis concerns the soul—or life, self-affection, moving oneself, seeing oneself—since the word *psyché* (the Greek term for soul, of course) means in French a large wardrobe mirror (MC 22 / 6). Yet the soul or the mirror necessarily includes a minuscule hiatus, *un écart infime*, around which the soul oscillates between seer and seen; otherwise, without the hiatus, the soul would be either pure activity or pure passivity, but not a relation *between* the self and the self, not the self *in* relation to the self. These prepositions designate the minuscule hiatus. And, just as necessarily, the minuscule hiatus is itself invisible: what I see is myself (as the subject of vision or as the object of vision), not the hiatus. The hiatus, therefore, remains necessarily a "blind spot" (*une tache aveugle*) (QRS 261 / 186; cf. MC 337 / 326), "this blind point . . . where our gaze steals away from ourselves at the moment we are gazing" (MC 20 / 4). The first major phase of Foucault's analysis locates the blind spot.

The second major phase of Foucault's analysis, which also takes place in the first section of the chapter, determines the emptiness or, more precisely, the invisibility or blindness of the empty square. In fact, the second phase shows that the mirror in the back of the painting does not double as representation or imagination do. The mirror does not double, *by means of resemblance*, something real and already visible. Instead, the mirror is a "metathesis of visibility"—this is Foucault's own phrase. The mirror produces a strange "chemical reaction," which *multiplies* the invisibility. Finally, the third phase, which takes place in the second section of the chapter (after the discussion of the artifice of the proper name), shows how the painting "diffracts" the multiplicity of the invisibility back into the painting. Here, in the third major phase, Foucault "unties" (the verb *dénouer* is one of his favorites) the mixture in which the invisibility consists. Foucault (following, we must assume, the painter's eye) differentiates in the indifference of the blind point. The result is that, despite the uncertainties produced by the invisibilities, the third phase of Foucault's analysis tells us who we might be. From the impossibility of vision comes the *peut-être*. The *maybe* is why the figure of man—he appears explicitly in the Velázquez painting—is possible in the modern epoch. Indeed, as we just mentioned, and as we shall see again at the end, this *peut-être* is why thinking in general is possible. Now let us turn to the three major phases of the analysis. Foucault locates the empty square by following the gaze of the painter.

The First Major Phase of the Analysis: Locating the Blind Spot

Foucault's starting point crystallizes the entire analysis that is about to unfold. The analysis starts with a retreat: "the painter is a bit *in retreat* from the picture [*légèrement en retrait du tableau*]" (MC 19/3, my emphasis).[20] A retreat is a repetition, the re-trait of the trait or — given the ambiguity of the French term *trait* — the repetition of the line that divides or partitions like a frame.[21] This retreat is a repetition that repeats emptiness, that repeats nothing visible, or that repeats invisibility. Thus the status of the repetition is uncertain. The painter, having thrown a glance (*un coup d'œil*[22]) toward his model, is immobile before he returns to paint the model's representation, either, Foucault says, "the last stroke or the first line" (*une dernière touche, le premier trait*): "perhaps," *peut-être*. The uncertainty of the "can be" is expressed by the "between," the *entre*. The painter is "between [*entre*]," Foucault says, "the fine point of the brush and the steely gaze." Between the brush, like a quill, for writing, and the eye for seeing, between spectacle and gaze, between words (written down, something to be read) and things (objects to be seen or gazed at). In this "between," according to Foucault, "the spectacle is going to free up its volume," like a book opening up its leaves, its recto and verso. Indeed, Foucault's analysis will divide the space of the painting in half, although he will remain focused on the sagittal line that emerges from the two sides.

The first sagittal line that Foucault analyzes emerges from the gaze of the painter.[23] Foucault says that "the painter gazes, he fixes an invisible point, but one which we, the spectators, can easily assign since this point is ourselves: our body, our face, our eyes" (MC 20/4). This sagittal line (*ce pointillé*) "reaches us infallibly and connects us to the representation of the picture" (MC 20/4). Now according to Foucault, only "in appearance" is this place simple, only in appearance does it look to be a place of "pure reciprocity." It seems that we are looking at the picture, out of which the painter, in turn, contemplates us. "*And yet* [et pourtant]," Foucault says, "this thin line of visibility that turns back on itself envelops a *complex* network of uncertainties, exchanges, and feints [*esquives*]" (MC 20/5, my emphasis). The painter is turning his eyes toward us, according to Foucault, only insofar as we find ourselves at the place of his "subject" (*motif*). We spectators are "in addition" (*en sus*) to the model. Although we are "welcomed" (*accueillis*) by the painter's gaze, we are chased away by it, since the painter's gaze is turned to the model. Yet, inversely, the painter's gaze accepts as many models as there are spectators, since it "is addressed *outside of itself toward an emptiness* [*hors du tableau au vide*]" (MC 20/4, my emphasis). The place where we are is the *empty square*, a "precise but *indifferent* place, where the one who gazes and the one gazed upon exchange themselves constantly" (MC 20–21/4–5, my emphasis).

The *indifference* of this place is the indifference of the mirror, and in a moment Foucault is going to speak of the mirror at the back of the room in terms of indifference (MC 23/7).[24] When I gaze at myself in the mirror gazing back, I cannot determine whether I am the one seeing or the one seen; in other words, I cannot differentiate, although there is a point, a hiatus, *un écart infime*, around which "subject and object, spectator and model reverse their roles to infinity" (MC 21/5). The reversal of roles, for Foucault, returns us to the canvas at the far left. If we could see the front side of the great canvas, we might be able to stabilize the exchange and make it determinate. But the canvas remains "*obstinately invisible*, it forbids that the relations of gazes ever be mappable [*repérable*] or definitively established" (MC 21/5, my emphasis). The canvas's "obstinate invisibility" forever makes the play of metamorphoses at the center between the spectator and the model unstable. This empty square where we are is indeed a place of pure and simple reciprocity *only in appearance. In fact*, "the painter," Foucault tells us, "actually fixes a place which, from instant to instant, does not stop to change content, form, face, identity" (MC 21/5). We do not *know*, in the sense of visible presence—we cannot see the front of the picture—who we are or what we are doing. Are we seeing or being seen?[25] The uncertainty of this question is produced by the fact that the canvas on the left always remains "obstinately invisible." As Foucault says, "at the moment when we are going to see ourselves transcribed by his hand *as in a mirror*, we can overtake nothing of the mirror but its lusterless back" (MC 22/6, my emphasis).[26] The great canvas, like all mirrors, includes a point of invisibility, a blind point (*un point aveugle*; MC 20/4).

The Second Major Phase of the Analysis: Multiplying the Invisibility

This comment is the first mention of a mirror in the analysis, and it marks a transition. Foucault draws our attention to the wall, which is the background of the painting—in particular to the mirror on the wall.[27] As is well known, Foucault specifies that the mirror in Veláz-quez differs from those found in Dutch genre paintings. In Dutch genre paintings, according to Foucault, the mirror doubles or "re-peats" what is already represented in the painting, but as in an "un-real space."[28] The mirror in Dutch genre paintings, in other words, functions as an imaginary doubling—the unreal space—of something real. Now the mirror in the Velázquez painting, as Foucault notes, indeed holds a "median position" on the wall and in the painting. And the line extending from it "partitions [the picture] in two" (*par-tage en deux*) from the top to the bottom. Therefore, it "could be" a perfect double of what is in the studio, and if the mirror did produce this perfect double, it would be of the same genre as the Dutch mir-rors. Yet it "makes nothing seen" of what the picture itself repre-sents. Instead, "its immobile gaze is going to seize out in front of the picture, in this region necessarily invisible which forms the external side of the picture, the characters that are disposed there" (MC 23 / 7–8). In other words, "it restores visibility to what remains outside of every gaze" (MC 23 / 8). What is the nature of this restitution that the mirror accomplishes?

Foucault tells us—and this comment is very important in order to understand the blind spot—the invisibility that the mirror "over-comes" (*surmonte*) is "not that of the hidden." The mirror does not "bend around [*contourne*] an obstacle," and it does not "turn over [*dé-tourne*] a perspective" (MC 23 / 8). What Foucault is implying here is that the invisible is never anything hidden like a secret, and thus it is not something that is imminently or potentially visible. The mirror is not making visible something that we could see from another per-spective. The invisible in Foucault remains "obstinately invisible." But there is more. The mirror is indeed presenting, making visible, allowing us to see something. As Foucault says, "At the back of the room, ignored by everyone, the unexpected mirror illuminates the figures that the painter (the painter in his represented reality, objec-tive, the painter at work) is gazing upon, but also really the figures who are gazing upon the painter (in this material reality that the lines and the colors have deposited upon the canvas)" (MC 24 / 8).

Both of these invisible figures—the ones gazed upon and the ones gazing—are, according to Foucault, "inaccessible." The first of these figures, the one that the painter is gazing upon, is invisible by *a com-positional effect* of this picture; here, Foucault is speaking of the *struc-ture* of the picture. The painter does not represent *inside* the composition of the painting what he is gazing upon. The second of these figures, the one who is gazing at the painter, is invisible by the *law* that presides over the very *existence* of every painting in general. In other words, we cannot see the figures that are gazing at the painter because the bottom of the painting—because the very being of all paintings—never extends far enough down to include in the painting the ones who gaze at it: you and me, us. Here, according to Foucault, the play of representation consists in leading one "form" of invisibility, the invisibility by the law of the existence of painting, to the place of the other, the invisibility by compositional structure "in an unstable super-position." The play of the representation, in other words, refers the two forms of invisibility immediately to the other extremity of the picture, to the back wall, to the mirror. But the mirror represents nothing shown within the space of the painting; in fact, it represents nothing visible at all. Instead, the mirror then guarantees a "metathesis of visibility." This metathesis is what the mirror is doing, according to Foucault. The word *metathesis* means a chemical reaction between two substances that produces two new substances. So, the mirror reacts, so to speak, "chemically" with visibility, in order to produce two invisibilities. The mirror, in other words, is not just the reflection of one invisibility; rather, it reflects what is "two times necessarily invisible from the picture" (MC 24 / 8). Reversing Pachero's advice to Velázquez, as Foucault says, what is outside the picture, what is outside the frame, *le cadre*, the *quadrum*, "the empty square must emerge from the image," and it emerges doubled or even multiplied. Therefore, we should recall, once again, what Foucault says about fiction in his essay on Blanchot: "fiction consists *not* in making us see the invisible *but* in making us see *how much* the invisibility of the visible is invisible."

The Third Major Phase of the Analysis: The Diffraction of the Multiplicity into Singularities

So far, in the first two major phases, we have been stressing that Foucault's analysis bases itself on the sagittal lines emerging from the painting: on the one hand, the gaze of the painter; on the other, the line emerging from the mirror. There are two more sagittal lines. As is obvious from the painting, there is a man standing in the doorway adjacent to the mirror; "like the mirror, he fixes the other side of the scene" (MC 26 / 10); this is the gaze, Foucault says, of a "distant visitor" (MC 27 / 12). The fourth sagittal line is that of the gaze of the infanta in the middle of the picture, surrounded by the painter, courtesans, handmaidens, an animal, a dwarf, and a buffoon. Given her figure in the painting, we know, according to Foucault, that the infant is the principal theme of the composition, the very object of this painting. By describing the composition of the picture, laying out a Saint Andrew's cross (which crosses directly over the face of the infant) and a curve that seems to hollow itself out into a vase, Foucault shows that the mirror and the gaze of the infant form two superposed centers of the painting. There are then two very close sagittal lines emerging from the gaze of the infant and from the mirror, according to Foucault, which converge right where we are, where we are gazing from. But, as the mirror indicates, the sagittal lines go out to the place where the sovereigns are. The mirror, of course, tells us what the painting is made of: the sovereigns (*les souverains*). Because the sovereigns are external to the picture, withdrawn into essential invisibility, they order the whole layout of the picture to their gaze; they order the whole representation. They are thus the genuine center of the composition, to which the gaze of the infant and the image in the mirror are submitted.

Foucault calls this center "symbolically sovereign" (MC 30 / 14), meaning (as we noted on the basis of Deleuze's structuralism essay) that the center allows different series to communicate without unifying them. In other words, what Foucault is doing in this third phase is "untying" (*dénoué*) the reciprocal relation of seer and seen; he is *un*making the preposition *en* in the phrase *spectacle-en-regard* (MC 29 / 14). So, in this symbolically sovereign center that is outside of the picture, there is a triple superposition of gazes. First, there is the gaze of the model at the moment he or she is being painted; second, there is the gaze of the spectator who is contemplating the scene; and third, there is the gaze of the painter as he is composing the picture (the painter Velázquez painting *Las Meninas*). These three "gazing" functions are "mixed together" (*se confondent*[29]), according to Foucault, in a *point* external to the painting. The point is at once ideal and real, and that is why it is symbolic. The point is an ideal point, since it is the point from which the painting would have to be viewed; we can imagine the geometrical relations between the point and the surface of the painting. But Foucault stresses that the point is a real point as well, since the painter really had to stand there in order to paint the painting, and the model really had to stand there in order to approve the painting and then be its spectator. Because this point is the place from which the gaze emerges, it is not gazed upon, and it cannot be gazed upon, since if it were, it would no longer be the source of the gaze, the gazing, but the gazed upon. So, "in this very reality, the point cannot not be invisible." We must stress the force of this statement: "il ne peut pas ne pas être invisible" (MC 30 / 15). It is necessarily impossible that this point be visible; it must remain stubbornly invisible, nearly an absolute invisible.

"And yet," Foucault continues, "this reality is projected into the interior of the picture, projected and *diffracted* into three figures, which correspond to the three functions of this ideal and real point" (MC 30 / 15, my emphasis). The painting has transformed the obstinate invisibility of the external point into a teeming presence of individuals or singularities. The function of the painter's gaze when he is painting is present there on the left, the painter with his palette. (The painting is, of course, the self-portrait of the author of the painting.) The function of the spectator's gaze is present there near the center, the visitor who takes in the whole spectacle. And, finally, at the center in the mirror there is the model's gaze, the reflection of the king and the queen, immobile like models. As Foucault notes, the reflection of the king and queen in the mirror seems to restore what is missing (*manque*) from each gaze as they are shown in the painting (MC 30 / 15). To the painter is missing the model—the king and queen—when he is copying his represented double on the picture. To the king is missing his portrait, which is being made on the opposite side of the canvas, so that he cannot perceive where he is. Finally, to the spectator is missing the real center of the scene, which the gazes of those in the painting are looking at, since he has taken the place of the king by "effraction." The spectator, in other words, has broken into the space and stands where the royal couple would be, and thus he cannot see what everyone else in the picture is looking at. Yet "perhaps" (*peut-être*), as Foucault says, "this generosity of the mirror is being feigned; perhaps, it hides as much as it manifests" (MC 30 / 15).[30] The mirror could just as well reflect the anonymous spectator or author of the painting, since they occupy the place where the king and queen are (MC 30 / 15).

But the way in which the diffraction into singularities has appeared in the painting—the artist and the visitor are in the painting, and thus they cannot be in the mirror, while the king is in the glass and thus cannot be in the painting—indicates, according to Foucault, that the mirroring relation is "interrupted" (MC 31 / 16). The mirroring relation fails because it reflects only one gaze, that of the sovereign; the mirror can never, therefore, make fully visible the singular point from which the gazes diffract. The "pure goodness of the image" cannot offer "in full light" this point (MC 31 / 16). The obsti-nate invisibility of the point is why the sagittal lines, which traverse the depth of the picture, are incomplete; a part of their trajectory is "lacking" (*manque*) to them, as if they cannot see the object that their vision is seeking outside of the painting. The "gap" in their vision is due to the absence of the king, but, as Foucault stresses, this absence is an artifice of the painting. The king could have been represented there in the painting. But this artifice conceals a vacancy, which, Foucault says, "is immediate" (MC 31 / 16). The painter must be va-cant from their vision when he is composing, since he must be behind his canvas. The spectator must be vacant from their vision, since he must be behind the king; the spectator has broken into the place of the king, but the king will chase him away. Again, the singular point from which the gazes emerge outside the painting is precisely what all the gazes in the painting are gazing upon and precisely what they cannot see, since it is a blind point. As Foucault says, "perhaps, in this picture . . . the profound invisibility of what one sees [the painter cannot see his model when he is painting; the king cannot see his portrait, which is on the reverse side of the canvas; the spectator can-not see the king, whose place he has taken by breaking into the place of the king] is inseparable from the invisibility of the one who sees — despite the mirrors, the reflections, the imitations, the portraits" (MC 31 / 16). Again, this singular point cannot *not* be invisible, not quite absolute and yet also not imminent.

Conclusion: *Les Suivantes*

The impossibility for the image to offer either the painter, the model, or the spectator in full light returns us to a question that we encoun-tered earlier, in the first major phase of the analysis: Are we seers or the ones seen? Indeed, the impossibility returns us to a very Foucault-ian question: Who are we? The question of who we are, we saw, arises from the uncertainty, the indifference, the emptiness of this place or square. Now, Deleuze tells us, again in his structuralism essay, that we can never completely fill the empty square, and we can never let it be completely empty. What we can do is "accompany" the empty square and "follow" (*suit*) it (QRS 266 / 190).[31] We should note that the title that Foucault gives to this first chapter of *Words and Things* does not correspond to the title given to Velázquez's paint-ing, *Las Meninas*; Foucault's title does not even correspond to the French translation of the Spanish title, *Les Ménines*. The title of the chapter, "Les Suivantes," refers to the handmaidens located around the infanta in the middle of the painting. Yet taking the word literally, we see that *les suivantes* —or better, making the term more neutral, *les suivants*[32] —refers to anyone in the entourage of the royal family, anyone who "follows" the sovereign; the entourage then would include not only the handmaidens but also the man in the doorway, the painter, and even the dog or animal. The title of the Velázquez painting has a strange history; at one point it was catalogued as *The Family of Philip IV.*[33] We see now what Foucault is doing with the title of his chapter. By means of this title, Foucault is "burrowing out" and reversing or even denying what the painting seems to represent, its principal theme, the sovereigns or the infanta. The title of the chapter is not "Les Souverains," but "Les Suivantes." Through this "burrowing word"—a phrase I am appropriating from *This Is Not a Pipe* (CP 47–49 / 36–37) —Foucault is providing a sort of answer to the question of who we are. The painter's gaze reaches us unfailingly and connects us to the representation. It can welcome us into the place of the king, and then also it can chase us away, since the painter is gazing at the model. We can be the guest (cf. MC 319 / 308), but also we can be, if we play on the French word *hôte*, the host, in the sense of undergoing something, the sagittal line of the painter's gaze penetrating like an arrow, turning us into ghosts. Who are we? We cannot be the sovereigns; we might be the followers.[34]

In this attempt to answer the question of who we are with the phrase *les suivantes*, you can see, I hope, that we have moved from an impossibility to a possibility: we *cannot* be the sovereigns; we *might* be the followers. To conclude, I would like to trace out a little this transition, this movement, from impotence to potency. To trace this movement out, we must recognize that, when Foucault speaks of the superposed sagittal lines of the mirror and the infanta exiting the picture—a sagittal line going out to a point external to the picture, to a point that is invisible, to a blind spot—he is speaking (like Kant, of course) of a condition. But he is speaking not (yet) of a condition of possibility; it is a condition of impossibility, a condition of *impouvoir* or inability—the impossibility that the pure goodness of the image will ever offer in full light the master representing and the sovereign who is being represented. The mirror relation has been interrupted. Early in his analysis, Foucault states that the Velázquez painting consists in uncertainties (MC 20 / 4). And in *This Is Not a Pipe*, Foucault says that Magritte's operation of the calligram being unmade produces the feeling of uneasiness (CP 19 / 20). So, we might say that the impossibility of seeing, this blindness, produces in us spectators the *feeling* of uncertainty.[35]

In his histories, Foucault always focuses on *des épreuves*; the French word *épreuve*, of course, means not only "test" but also "experience."[36] The blinding experience of uncertainty affects me, subjects me, puts me to the test. Being put to the test, the experience is the experience of death in life, the experience of the impossible.[37] I *cannot* go over the limit of the blindness. I am in default (*défaut*) in relation to vision, to visible presence, and yet I feel the necessity of going over the limit (*il faut*). On the basis of the "impossible to determine," on the basis of the "impossible to differentiate," on the basis of the "impossible to see," a "maybe" arises, or, more precisely, a "can be" arises. Here we have the *source* of all the occasions of *peut-être* with which Foucault punctuates his analysis of the Velázquez painting. The condition of impossibility becomes the condition of possibility; this double condition is the law, the a priori law exhibited by *Las Meninas*. The feeling of impotence becomes the feeling of power.[38] Precisely because I cannot see *determinately*, it is possible that I am seeing this or that or that (*ceci*). In other words, precisely because I cannot see, I can see more. The painting, we recall, diffracts the blind point back into the picture, into the painter, into the sovereigns, and into the man in the doorway (who is there in "flesh and blood"; MC 26 / 11). The diffraction into this one or that one or that one, each singular, is possible only on the basis of the impossible point. The painting diffracts the blind point, we might say, into all the followers—*les suivantes*— around the infanta, who is virtually under erasure due to the Saint Andrew's cross. These followers, including the upright man, who is the man of "man and his doubles," including the jester and the animal,[39] the multiplicity of the followers might be (*peut-être*) what is possible in our uncertainty.[40] But when it diffracts the blind point, the eye can no longer see. Then, perhaps, the eye can be productive.[41] It can become a spring (*une source*), pouring out tears, or pouring out oil onto the canvas, or pouring out ink onto the page; the eye, per-haps—the blind, of course, "see" with their hands—comes then closer to the hand, to the hand of the surgeon, who is making an incision, making a difference, in a word, thinking, "a perilous act" (MC 339 / 328).

9

"This Is What We Must Not Do"
The Question of Death in Merleau-Ponty

Dedicated to the Memory of Martin C. "Mike" Dillon

What is a suicide bomber? In essence, the death of a suicide bomber is no different from any other suicide; the action is inconceivable without auto-affection. Indeed, as with all suicides, the auto-affection in which the suicide bomber engages is contradictory: he affects himself in order to end all his own auto-affection, all his own affectivity in general. Yet what distinguishes the suicide bomber from other suicides is that his actions end the lives of many others. The increase in destruction makes the contradiction more severe, even more paradoxical, at least for us in the West. In the East, the suicide bomber is a religious martyr (a *shahid*), who transcends life on earth for paradise.[1] The difference in names perhaps indicates a distinction in regimes of thought.[2] But that we Westerners find a "suicide bomber" paradoxical indicates that our epoch, as Foucault claimed, is an epoch of *bio-power*. For us, an increase in power does not flow from death; life must be managed, that is, preserved and enhanced. Now it seems to me that we can understand and transform something like our "global war on terror" only through a re-engagement, or better, through a new engagement with the concept of life. Here is one reason, a political one, for renewing the concept of life, but there is another, a philosophical reason. I have just used a formula that comes from Heidegger: "preservation and enhancement of life."[3] Despite his immense and unquestionable importance, however, Deleuze and Derrida have criticized Heidegger's thought of being for its inability to think genuine singularity, an inability that also implies that Heidegger's thought of being cannot think genuine multiplicity. If we cannot think singularity and multiplicity genuinely, that is, in a non-representational way, can we really say that we have overcome *metaphysics*? The final metaphysics, of course, that Heidegger considers is that of Nietzsche, the metaphysics of the will to power, the very metaphysics that supports bio-power. Only a concept of life reconceived in terms of *powerlessness* allows us to think singularity and multiplicity in a genuinely nonmetaphysical way. So it seems to me that, at this moment, for these two reasons—one political, the other philosophical—we must develop something like a neo-vitalism.

Here I would like to ask how Merleau-Ponty's thought might contribute to a new concept of life. Starting at the end of the nineteenth century, scientific and philosophical investigations of life have been divided between two concepts, the concept of lived-experience (*Erlebnis* in German, *vécu* in French) and the concept of the living being (*Lebendige* in German, *vivant* in French).[4] Merleau-Ponty transformed the phenomenological concept of *Erlebnis*, understood as a strictly subjective and internal experience, into an experience that is always ambiguous. But he also transformed the concept of the living, in the nature lectures presented in the late fifties.[5] Conceived as a "propaedeutic" to the ontology of the visible and the invisible (N 265/204), the nature lectures engage in an "archeology" (N 335/268) of the concept of nature.[6] By undoing the constructions of classical thought and of modern science, Merleau-Ponty intends to "return to dynamism" (N 23/7), "dynamism" in the literal sense, *dynamis*, potentiality.[7] As we shall see, potentiality (or power) works this way in Merleau-Ponty's archeology of nature: because nature is *hollowed out*, it *carries* future developments. For Merleau-Ponty, we might say, a lack of power makes power possible. As with lived-experience, the idea of power in Merleau-Ponty, the power of the living, is ambiguous; the *archē* is mixed with powerlessness.[8] Yet if the future is already there, if an organism "has in potential what its complete life will be in the future," if its last "note," as in a melody, is already there at the very beginning, does not this last moment of life imply that *death* too is already there? Merleau-Ponty, we know, has no discourse of death anywhere in his writings.[9] This is the question I would like to address: Why does Merleau-Ponty never make death, in any form—one's own death, the death of the other, prepersonal death—a kind of theme? I will try to answer this question in three steps.

First, we shall approach an answer to the question of death by examining what Merleau-Ponty has to say about God. Perhaps indicative of his time, Merleau-Ponty takes up the idea, found in, to use his words, "contemporary philosophy," that God is dead. The pure nihilism, as we shall see, of contemporary philosophy complements the pure positivism of what Merleau-Ponty, in his 1955 "Everywhere and Nowhere," calls "great rationalism" (*le grand rationalisme*), which refers to Classical thought, the epoch from Descartes to Kant. For Merleau-Ponty, veering off into either of these two complementary poles is precisely what we must not do if we want to conceive the principle of nature. Therefore, in a second section, we shall attempt to determine as precisely as possible the *archē*, the origin or principle, that Merleau-Ponty's archeology attempts to retrieve. *Principle* (*principe*) is one of the most frequent words in the nature courses. Merleau-Ponty provides for us a number of imperatives by means of which to conceive the principle of nature. But his own way of obeying the imperatives amounts to a "translation" of the "secret" of great rationalism. The translation will show that the "barbaric principle" of nature—Merleau-Ponty borrows this phrase from Schelling—is in fact *tranquil*. It is this tranquility, I shall argue, that keeps Merleau-Ponty from making death into a theme. In a concluding section, I shall turn to Merleau-Ponty's one comment concerning Xavier Bichat's definition of life as the set of functions that resist death. This definition will return us to the idea of life and war. But first let us talk about God. After all, Merleau-Ponty says in the discussion following his presentation, in 1951, entitled "Man and Adversity" that "for me, philosophy consists in giving another name to what has long been crystallized under the name of God."[10]

Pure Positivism and Pure Negativism

In the first nature lecture in particular (dating from the academic year 1956–57), Merleau-Ponty takes up the idea of God, because the Classical concept of nature "has not stopped explaining us" (N 25/8). Our thought of nature and our modern science are still based on what Descartes initiated. Because of the relation between us and the Classical concept of nature, Merleau-Ponty says, in "Everywhere and Nowhere," that "great rationalism" is "close to us" (S 187/149).[11] Indeed, he calls for a contemporary "translation" of great rationalism. This is what he says: "Descartes said that God is conceived of but not understood by us, and this 'not' expressed a privation and a defect in us. The modern Cartesian *translates* [my emphasis]: the infinite is as much *absence* as *presence* [Merleau-Ponty's emphasis], which makes the negative and the human enter into the definition of God" (S 189 / 150).[12] What must be translated in great rationalism for Merleau-Ponty is the idea of God as a *positive infinite*; this idea, he says, is "the secret" of great rationalism. What attracts Merleau-Ponty to the idea of God as the positive infinite is that it consists in a part-whole relation. As Merleau-Ponty says in "Everywhere and Nowhere," "every partial being directly or indirectly presupposes [the positive infinite] and is in return really or eminently contained in it" (S 187 / 149). The comment means that every part is smaller than or relative to the whole, which is infinite or larger. Because the whole is *larger*, *le grand rationalisme* is in fact "large rationalism." But in the contemporary translation, the finite will enter into God. The translation then must take up the idea of a God that can die. Now, I think that this translation of the idea of God as the positive infinite into a kind of infinite finitude is precisely the project that Merleau-Ponty undertakes in his later philosophy; elsewhere I have argued that Merleau-Ponty's last published text, "Eye and Mind," attempts precisely to translate the positive infinite.[13] I am going to take up the translation once more in a moment. But first we must examine the two ways in which the positive infinite functions in the Classical epoch: Cartesianism and Spinozism. Then we must consider "the philosophy of God is dead." Does nihilism translate the "secret" of large rationalism? The answer to this question will of course be "no." In any case, like Merleau-Ponty, let us start with Cartesianism.

The idea of God determines the Cartesian conception of nature. In Descartes, nature is defined as a machine, a definition, Merleau-Ponty tells us, that "blends together a mechanism and an artificial-ism" (N 27 / 10). Conceived as a machine or artifact, nature requires an artisan, God, who builds the machine. As in human production, where the artisan is separated from his or her product, nature is sepa-rated from God.[14] Nature in Descartes—and for us, according to Merleau-Ponty—is conceived as a "pure exteriority," outside of or separated from the principle of its creation, God. Interiority (which means intentional creation, teleology or finality) withdraws into God. Yet in order to make the meaning of nature evolve into pure exterior-ity, Cartesian ontology, according to Merleau-Ponty, had to engage in a reflection that "purifies" nature of what resists the faculty of the understanding. Thereby it is able to take a step toward essence (N 170/126). Not only is nature pure exteriority in Classical ontology (that is, separation), but also it is pure extension (N 170/126). Nature is reduced to the "geometral" or the surveying plan (N 64/40; OE 10/122; S 30/22). Nature is "spread out" (*partes extra partes*) as a pure object before the pure understanding (cf. N 153/113). The formulas for Cartesian dualism are well known. The pure object over and against a pure subject and the pure exterior over and against the pure interior are real distinctions, which are also separations. But for Merleau-Ponty in the first nature course, the dualism itself seems to divide the six Meditations in two. Unlike the Descartes of Meditations I to III, who attempts to idealize nature as an essence, the Descartes of Meditations IV to VI takes into account the pressure of the actual world (N 35/16). The pressure results in the fact that Descartes also speaks of *actual* extension alongside the ideal extension of *partes extra partes*. Here in the composite union of my soul and my body, there is obscurity; human understanding is finite. The finitude of human understanding, according to Merleau-Ponty in "Everywhere and Nowhere," is a "defect" (*défaut*) (S 189/15). It is through this defect or lack that the second Descartes (the Descartes of Meditations IV through VI) understands the relation of parts or creatures to the whole or God. For the second Descartes, there is not a dualism but a "mixture." Indeed, as we have been trying to show, all of Merleau-Ponty's thought could be defined as a mixturism. Now, it is precisely this mixture that Merleau-Ponty *cannot* find in Spinoza.

Spinozism, therefore, for Merleau-Ponty is more of an error than Cartesian dualism. Spinoza, according to Merleau-Ponty, knows nothing of the "opposition between real extension and extension in thought" (N 34/15). Intelligible space and perceived space in Spinoza "are separated only by a difference of more or less finite ideation" (N 34/15). Because the relation is only a question of more or less ideality, there is no mechanism in Spinoza. The Cartesian separation therefore is eliminated. God again is a positive infinite, essence itself (N 30/13). He is self-causing, being "through" itself, engendering all of itself and all of nature (N 60/37). God is the first in the *hierarchy*; the supreme being is prior to nature in a *linear relation*. According to Merleau-Ponty, in Spinozism it is not possible to conceive another being and another world: God is a necessary being, with the result that "finitude is nothing other than this drawing of the finite from out of this power [*puissance*] of an infinite being" (N 60/37). The finite is drawn out of the infinite, of course, by means of negation. The well-known formula of Spinozism "All determination is negation" means, according to Merleau-Ponty, that negation "is *only* negation, [*only*] irreality in relation to the supreme reality" (N 208/156, my emphasis). In Spinozism, the finite parts no longer relate to the infinite God as defects or lacks but rather are contained in God "analytically." Spinozism, therefore, is a pure positivism.

The errors of Cartesian dualism and Spinozistic monism define the Classical epoch. But in "contemporary philosophy" there seems to be a still more serious error for Merleau-Ponty. This error is nihilism, which Merleau-Ponty discusses in the context of a reflection on Heidegger. Merleau-Ponty insists that Heidegger's thought is *not* nihilism; it is defined neither by "the philosophy of nothingness" (the allusion to Sartre is obvious) nor by "the philosophy of God is dead" (Merleau-Ponty mentions Nietzsche; NdC 1959–61, 119, 146). Using a formula from Heidegger, Merleau-Ponty says that, in nihilism, God's death is "a concealment that conceals itself" (NdC 1959–61, 119); in other words, God withdraws entirely. The complete death of God means that God has separated himself from being, withdrawing into complete nothingness, having no effect on the present. This formula implies that God's death is a pure nihilism. Clearly, Merleau-Ponty rejects this idea of pure nihilism as a false nihilism. It does not fulfill the project of a *translation* of the idea of God understood as the positive infinite into a finite infinite. The pure nihilism of contemporary philosophy merely *reverses* the pure positivism of large rationalism. Instead of a positive infinite, we have a negative finite; instead of two planes in Descartes (one real, one ideal) or one plane, as in Spinoza (which, in Merleau-Ponty's reading, is completely ideal), we have one plane in contemporary nihilism that is completely actual. Indeed, the "actualism" of contemporary philosophy parallels modern science (N 344/276), which, in "Everywhere and Nowhere," Merleau-Ponty calls "small rationalism" (S 186/148). Modern science is small because its operations function independently of anything larger: God in modern science is *merely* dead. But just as the positive infinite evolved from the Judeo-Christian tradition, in fact from the biblical God, so contemporary nihilism evolves from the Christian God.[15] Merleau-Ponty quotes Jesus' last words: "Why have you abandoned me?" (N 180/134). Therefore, rejecting both pure positivism and pure negativism, Merleau-Ponty thinks that "we must grasp God as the keystone; that is, he is what the *edifice* supposes and what makes the Whole stay together. It is this paradoxical relation that we must look in the face" (N 180 / 134; Merleau-Ponty's upper case, my emphasis).

The Tranquil Principle of Merleau-Ponty's Archeology of Nature

What Merleau-Ponty here is calling "the paradoxical relation" is the principle of nature, whose consideration, we recall, is the "propae-deutic" for *The Visible and the Invisible*. In the nature courses, Merleau-Ponty gives us an imperative for conceiving the principle of nature. It seems to me that even today, fifty years later, we must obey this imperative if we want to conceive an *archē*, an origin or a princi-ple. Here is the imperative in its negative form. The principle must be conceived neither as positive nor as negative, neither as infinite nor as finite, neither as internal nor as external, neither as objective nor as subjective; it can be thought neither through idealism nor through realism, neither through finalism (or teleology) nor through mechanism, neither through determinism nor through indetermin-ism, neither through humanism nor through naturalism, neither through metaphysics nor through physics. Veering off into one of these extremes is precisely "what we must not do" (N 203 / 152). In short, for Merleau-Ponty — but this is something Deleuze says too in his 1968 book on Spinoza[16] — there must be no *separation* between the two poles. But also, there must be no *coincidence*. Neither Platonism (separation) nor Aristotelianism (coincidence) is adequate (N 206 / 155). Merleau-Ponty recognizes, in fact, that separation and coinci-dence implicate one another just as the pure positivism of the Classi-cal epoch implicates the pure nihilism of contemporary philosophy: pure positivism gives us pure negativism by means of positivism's mere negation (N 96 / 65). The positive formula for Merleau-Ponty's imperative would be the following: instead of either a separation or a coincidence, there must be "a hiatus," *un écart* (N 208 / 157), which *mixes* the two together (N 164 / 121). The mixture implies that, in fact, we are not really dealing with a principle, at least in its most traditional formulas: indivisible, sovereign, unified, a one. It is a prin-ciple without principle.

Merleau-Ponty gives us a specific way of conceiving the "princi-ple" (at this moment the scare quotes are necessary) of the mixture that defines nature. In *The Visible and the Invisible*, the ontological name for the principle of nature is the "jointure" of the visible and the invisible (VI 154 / 114). What follows is a reconstruction of the jointure, based primarily on the nature lectures. Merleau-Ponty conceives the principle—that is, God—as an abyss (an *Abgrund*) or as a nonbeing (N 60 / 37). That God is nonbeing means that He has died. The idea of an *Abgrund* that Merleau-Ponty appropriates comes not only from Schelling but also from Heidegger. So we can rephrase our first formula this way: God is concealment, but not the kind of concealment that we saw in pure nihilism. There is, as Merleau-Ponty says, a "true nihilism," which is based on "the true nothingness" (NdC 1959–61, 137). The true nothingness is a kind of concealment that does *not* dissimulate itself (NdC 1959–61, 146). The nondissimulated concealment continues to have effects in what is unconcealed. In His withdrawal, God is still present, and therefore we must say that He is still alive. We can see an answer to the question of death in Merleau-Ponty approaching. But the answer will become determinate only if we continue to reconstruct his idea of a principle. As we anticipated, Merleau-Ponty's idea of a principle translates large rationalism's idea of the positive infinite. It seems to me that Merleau-Ponty's translation takes place in two more phases.

First, God must be conceived as a nonbeing who continues to have effects in the present. Then, second, the positive infinite must be transformed from an infinite of essence into an infinite of existence (N 60 / 37).[17] Being an infinite of existence, the infinite is mixed in with the finite. Therefore, unlike both Classical positivism and contemporary nihilism, no longer is there only one ideal or essential plane, no longer is there only one actual or factual plane. And yet there is no dualism, no two planes (cf. N 104 / 72).[18] For Merleau-Ponty, not being a dualism, the mixture of infinite and finite means that there is no longer a hierarchy; there is no linear order between the infinite and the finite, no priority of the infinite over the finite (N 31 / 13). Rather, as Merleau-Ponty says, instead of a hierarchy, there is an "architecture" or an "architectonic."[19] The image of an architectonic refers us back to the image of God as an "edifice," an image that appears already in "Indirect Language and the Voices of Silence" in 1952 (S 50 / 39). In "Indirect Language," he tells us that the stones of an arch "shoulder one another," that is, there is a relation of mutual dependence between them, without which the arch would fall. The relation between the stones (called "diacritical difference" in "Indirect Language") makes the arch stand and bear weight; the relation itself is the keystone. This image shows, because literally there is *nothing* between the stones in the arch, that, in Merleau-Ponty, we do not have a positive principle (N 203 / 152). Indeed, at one point in the nature lectures he speaks of a "negative principle"; at another, he speaks of a "finite principle" (N 208/156). Nature, in other words, is not "all-powerful" (N 208/156); we no longer have the God of large rationalism but rather a "finite God" (N 81/53).[20] There is a kind of "weakness" here that Merleau-Ponty designates with a number of terms (N 31/14): evil (N 65/41), weight (N 74/48), resistance (N 117/83), obstacle (N 91/61), and residue (N 117/83); in a word, "contingency," which Merleau-Ponty says "must certainly be at the heart of our thought" (N 54/33). It is this contingency, found in the union of the body and the soul, that the understanding, in Descartes or more generally in large rationalism, could not grasp. We must stress this point again. The first phase of Merleau-Ponty's translation is that the principle of nature is nonbeing; the second is that the principle becomes existential, a transformation that turns the principle into one of "fault" or "defect" (*défaut*) (N 99/68). Having become an infinite of existence, nature is not the almighty; it suffers from powerlessness, and therefore from death. Following Schelling, Merleau-Ponty therefore calls the principle "barbaric."[21]

The third phase of Merleau-Ponty's translation, it seems to me, however, overturns the barbarism. Even though there is no separation between the finite and infinite, the finite is not related to the infinite as an analytic negation. As Merleau-Ponty says, "there is a dignity to the positive finite" (N 61/37). This comment does *not* mean that now Merleau-Ponty is constructing a positive principle; there is still the in-finite, with its negative prefix. Indeed, Merleau-Ponty defines the existential (not essential) relation between the finite and the infinite as a *"fecund* contradiction" (N 61/37, my emphasis).[22] The important word here is *féconde*, fruitful. The barbaric principle is becoming the fruitful principle. The fecundity of the contradiction comes from the fact that the mixture of the infinite into the finite also mixes the eternal into time. Or, more precisely, now following Whitehead, Merleau-Ponty's principle is one of "process," an English word that Merleau-Ponty renders in French as *passage* (N 162/119). The word *passage* indicates that the principle of nature is the past, the past passing into the future. Whitehead had called process a relation of "overlapping," and Merleau-Ponty translates this English word with the French *empiètement* (N 157/115). This term appears frequently in Merleau-Ponty's last writings, rendered in the English translations usually as "encroachment." Nature is the past that has passed but that is also still there, encroaching on the future, giving the present a kind of thickness and depth. Nature, therefore, in Merleau-Ponty's interpretation of Whitehead, is a kind of memory: "the memory of the world" (N 163 / 120). How are we to understand this memory?

What makes memory come to life, what makes the past bear fruit (again, its fecundity), is not something positive. Merleau-Ponty wants to eliminate the idea of the possible as "the simple preformed reservoir," and yet, he says, "it is not true that everything is actual" (N 306 / 241). Because God is concealment, or because "nature loves to hide" (N 325 / 258, NdC 1959–61, 100), what is possible in nature is "lacking" (*un manque*). The present passes, and yet the past is still there "in relief" (*en creux*). Merleau-Ponty therefore speaks of the lack that defines the past as "a hollow" (*un creux*).[23] The passage of nature, as we see it in evolution for example, is for Merleau-Ponty a process of hollowing out. By hollowing itself out, nature institutes a limit within itself. Although Merleau-Ponty never says this, perhaps here in this hollow nature already in itself contains art, implying that nature is never pure, never noncultural. What Merleau-Ponty says is that the hollowing process makes a "*scission* in the infinite that produces the finite" (N 61 / 37, my emphasis). As we have seen, Merleau-Ponty uses contradiction to define the scission of the infinite and the finite, but it seems to me that he ultimately makes use of the Leibnizian idea of incompossibles.[24] Being modalities of one subject, incompossibles amount to a kind of contradictory identity, or, as Merleau-Ponty says, a "non-difference with itself" (N 207 / 156). It is possible for Adam to sin or not to sin, but both possibles cannot be actualized at the same time. And even if this possible is actualized, the other possible is still there in potential. But we know that the Leibnizian concept of incompossibility implies that it is *not* simply the case that *one* possible individual is in the process of being actualized; in passage, a whole world is coming into actuality, the world in which Adam sinned. This sinful world consists in *a set of possibles* that diverges from other sets that remain possible. In other words, using Merleau-Ponty's own terms, nature hollowing itself out outlines a way of "measuring" or a "dimension."[25] Nature hollowing itself out is dimensionality. The limit that is "sketched" is not a black line (N 201–2 / 151); it is gray and therefore indeterminate (an indeterminate set of possibles). The "outline" gives only a *sense* of what is missing and what is necessary for it to be filled in.

But, in Merleau-Ponty, we must *not* be misled by the word *scission*. What had been potential in nature was not something possible that needed only to be realized. Instead, what was possible was missing and yet was present in relief (*en creux*), like a mirror image that presents a, so to speak, "double." The doubling implies that "resemblance is the operation of Nature" (N 242 / 185). The operation of mimicry between an organism and its environment or between one animal and another implies, according to Merleau-Ponty, that there is no "rupture" in the passage (N 65 / 41). As Merleau-Ponty says, "there is here more of a sliding [*glissement*] than a rupture" (N 209 / 155). Merleau-Ponty conceives passage as continuity: "Nature is something that *continues*, that is never grasped in its beginning, although appearing always new to us" (N 160 / 118, my emphasis).

Like Bergson, Merleau-Ponty is a continuist.[26] Yet we see here a difference from Bergson: "although *appearing* always new to us" (my emphasis). For Bergson, what *appears* to us (the appearance as opposed to what is real) is something that looks to have merely realized a possibility that was already given; the forms of life resemble each other, but this resemblance, for Bergson, is an illusion. Bergson calls this the "retrospective illusion," the illusion in which "the whole is already given" at the beginning. In the nature lectures, Merleau-Ponty repeatedly refers to this illusion. Indeed, we can see him struggling with it in the passage I just quoted: nature "is never grasped in its beginning." Yet Merleau-Ponty does not completely accept Bergson's idea of novelty when he raises these questions: "Does this critique of the bad retrospective always liquidate the idea of the possible? If we take away the fictive possible, must we reduce Being to the actual?" (N 100 / 69). For Merleau-Ponty, therefore, there are no "discontinuous explosions, in the duration, in life, in history" (N 100 / 69). And when we retrospect, what we recognize is the absence of what had later come into existence. For Merleau-Ponty, this recognition is true. While for Bergson the different forms of life, the limits between the forms of life, break continuity (and only the dissolution of the forms re-establishes continuity), for Merleau-Ponty, the forms resemble each other: "the forms have a relation of sense among them" (N 208 / 157).

In the third nature course, called "Nature and Logos: The Human Body" (dating from the academic year 1959–60), for example, we find a series of "sketches" of the human body (because, as we saw, nature hollowed out sketches; we must recognize that *esquisse* is a technical term in the nature lectures), all of which aim to grasp humanity as a way of being a body, at its point of emergence in nature (N 269 / 208). These sketches place the human body's arrangement in the process of evolution. The morphological variation from the ape—that is, the passage from one form to another—Merleau-Ponty says, is "minuscule" (*infime*): the human body is bipedal; being upright allows the hands to take over from the jaws the function of prehension; in turn, the freeing of the jaw allows the muscles of the head to relax; the brain expands while the size of the face diminishes, which allows the eyes to come closer together in order to focus on what the hands grasp (N 334/267). With the mention of eyes and hands, we are very close to Merleau-Ponty's idea of the flesh, and, of course, in the third nature course he speaks frequently of the flesh. But the point I want to make with this quotation is that, in this way, "man"—here Merleau-Ponty is quoting de Chardin—"comes silently into the world" (N 339/272, N 334/267). For Merleau-Ponty, "there is *not* a decisive cut [*coupure*] between the stone and the animal or between the animal and man" (N 105/73). The lack of discontinuity is why Merleau-Ponty repeatedly speaks of our "kinship" (*parenté*) with the animals (N 335/268, NdC 1959–61, 148). Because of this kinship, nature, for Merleau-Ponty, is the "mother," the *Urmutter* (N 46/26), the womb, hollowed out. And yet, in being hollow, nature is the "cradle" (N 284/222), capable of "carrying" us (N 163/120). Continuing the image of carrying (the French verb is *porter*[27]), Merleau-Ponty also says that nature is "Noah's ark," carrying us on the water, or it is "the soil of the earth" (N 110–11/77).[28] Like the earth, nature has a kind of "inertia"[29] or "solidity" to it (N 316/249) that implies that it consists in a kind of "permanence" (N 158/116) or "eternal return" (N 20/4). Like the pyramids, nature, in Merleau-Ponty, never really dies; it never really forgets.

Conclusion: Tranquility and Conflict

Nature's never forgetting means that Merleau-Ponty's archeology of nature is in fact a memory of memory. Nevertheless, Merleau-Ponty's thought is always a "primacy of perception." Even though, as we just saw, following Bergson, Merleau-Ponty in the nature lectures struggles against "retrospective illusions,"[30] against the illusions that later developments, say, in evolution, were already positively present in the past as a preformed possibility, he still conceives the past as a mode of the present; he still conceives memory as a variant of perception (VI 248/194).[31] As a result, forgetfulness, the powerlessness of memory, in Merleau-Ponty is an "imperception" (VI 250/197). As he says in the 1955 passivity lectures, repeating a formula for memory that comes from Aristotle's *Posterior Analytics* (100a10–15), forgetfulness "counts for consciousness as a soldier counts for his company"; forgetfulness is a "secret memory."[32] In "Indirect Language and the Voices of Silence," Merleau-Ponty even says that the tradition of painting is "*the power to forget origins* [Merleau-Ponty's emphasis] and to give to the past not a survival, which is the hypocritical form of forgetfulness, but a new *life* [my emphasis], which is the noble form of memory" (S 74/59).

Given what we see in the nature lectures, it is possible to claim that the orientation of Merleau-Ponty's thinking was already set in *The Structure of Behavior*. There, already in 1942, he claims that death must *not* be "deprived of sense" (SC 240 and 220/223 and 204). We can immediately confirm that this orientation continues into his final writings. In *The Visible and the Invisible*, Merleau-Ponty says, psychoanalysis "is destined . . . to transform the powers of death into poetic productivity" (VI 155/116). In other words, death, or "Thanatos" (to use the psychoanalytic term from Freud), is meaningful (N 288/226);[33] it continues the dynamism; it continues the "active becoming" of sense.[34] No matter what obstacle life encounters, the obstacle takes on a "positive value" (N 91/61). For Merleau-Ponty, therefore, death is no different than a mutation; it is merely a metamorphosis that does not cut apart the kinship with what came before and with what will come later. All the ones who have already died, for Merleau-Ponty, all those who are buried in the earth *carry* us and carry all of those who will come in the future. Even in its forgetfulness, tradition is pregnant, carrying the future.[35] For Merleau-Ponty, death, therefore, is really about giving birth. Indeed, we could see this view on the very first page of the first nature course: Merleau-Ponty tells us that the Latin word "nature" comes from *nascor*, which means not "to die" but "to be born" (*naître* in French has the same Latin root; N 18/3).

Because nature really concerns birth, Merleau-Ponty concludes his examination of animal behavior, in the second nature course, by rejecting the definition of life that Bichat invented at the beginning of the nineteenth century: life as the set of functions that resist death (N 248/190). For Merleau-Ponty, this definition (which will reappear without being identified as such in chapter 3 of *The Visible and the Invisible*; VI 117/84–85) implies that life concerns itself only with survival; it implies, for Merleau-Ponty, "Darwinism" (N 230/175). Nevertheless, it must be recalled that Foucault, in *The Birth of the Clinic* (in 1963), will praise Bichat for having invented, through this very definition, a new form of vitalism, one based on "mortalism."[36] Following Bichat's definition, Foucault conceives death as being internal to life (still no separation), turning life into a *conflict*. Unlike Foucault, Merleau-Ponty, it seems, does not see that Bichat's conception of life implies not only that being is always "out of joint" (there is no "jointure" or balance), but also that knowledge itself (medical knowledge, for example), even a surplus of knowledge, is produced in conflict. Here it seems we have to confirm what many commentators have said over the last couple of decades: Merleau-Ponty's thought is based on a kind of tranquility. What has become of the barbaric principle, the wildness or savageness, the evil of nature? It has become tranquil, "the true tranquility," as Merleau-Ponty says in the first nature course (N 79 / 52).

Nevertheless, it seems to me that a slight change in emphasis in how we conceive the minuscule hiatus is able to move us from Merleau-Ponty's tranquility to Foucault's bellicosity. In fact, I think that it is *not* possible to get to Foucault except through Merleau-Ponty's translation of the "secret" of large rationalism. That is why a moment ago I said that Merleau-Ponty's imperative—"This is what we must not do"—must still command our thinking. We must recognize, however, that our "today" is no longer the same as Merleau-Ponty's. What has happened? Since Marx has become a specter, signs have changed. We see more distinctly that we live in an epoch of bio-power. Probably, analytic philosophy's interest in naturalism participates in bio-power's will to knowledge (which aims at the preservation and enhancement of life). If we are going to resist bio-power's will to knowledge, we must construct a counter-discourse, a counter-knowledge. I think that we can call it "life-ism" or even "neo-vitalism." Yet, in the construction of life-ism, we cannot reproduce all the problems associated with the vitalism of the nineteenth century, problems clearly indicated by Heidegger in his lectures on Nietzsche. We require, as Merleau-Ponty saw, an archeology of nature, but one that aims not at dynamism but at powerlessness. On the basis of *this* archeology, powerlessness would not be a modification of power. We must conceive death in connection with life, but in such a way that death produces discontinuity in life, that it is an "unfold" in the folding together. While in Merleau-Ponty death is more about birth, we must think that birth is more about death. Is it not the case that as soon as we are born we are ready to die? This death in life would be a teeming presence. It would remain, we might say, non-sensical.[37] *This* non-sense would give us no direction; it would not carry us, especially it would not carry us like our mother. This non-sense would be a beast. All that we can do is follow it this way and then that way, accompany it. Yet, although sometimes we would follow it as though it were an animal that we are tracking, at other times, the beast would turn and follow us.[38] Here we do not have tranquility but conflict.

Tranquility and conflict, carrying and following,[39] perhaps only this difference (between the Latin *ferre* and the Latin *sequere*) can allow us to distinguish between Merleau-Ponty's thinking and that of Foucault. Perhaps it is the only way to start to determine the difference, as well, in their political thought. Here is the difference in its most reduced form. In his 1955 *Adventures of the Dialectic*, Merleau-Ponty claims that "politics begins with accepting mediations."[40] Barely twenty years later, in *Discipline and Punish*, Foucault claims that "Politics is the continuation of war."[41] Foucault's reversal of Clausewitz's saying allows us to return to the problem with which we started. The global war on terror is being waged for bio-power. And even if the difference in the names by means of which a person who kills himself in order to kill others—"suicide bomber" versus "martyr"—indicates that the East may not be located within a regime of bio-power but is rather in a regime of sovereign power, we must recognize that even sovereign power concerns the power to take life, the power of death. No matter what, therefore, we must re-engage the concept of life. That means we must try to conceive the minuscule hiatus that opens up the possibility of auto-affection, even the auto-affection by means of which one kills one's own self in order to kill others.

10

Metaphysics and Powerlessness
An Introduction to the Concept of Life-ism

Sum moribundus
— Martin Heidegger, *The History of the Concept of Time*

Contemporary politics finds itself, as Foucault showed in his 1976 *The History of Sexuality, Volume I*, within a regime of *bio-power*.[1] This regime, on the one hand, allows a deadly epidemic to develop in one population—I am thinking of the AIDS epidemic in Africa—while on the other it allows the life of one paralyzed individual to be preserved almost indefinitely, as in the 2005 case of Terri Schiavo in the United States.[2] It seems to me that we can understand this contradiction between a population and an individual only by means of a renewal of the concept of life. The philosophical reason for a renewal of the concept of life comes from a worry about Heidegger's thought, a thought that undoubtedly still overshadows all contemporary thinking. As we have pointed out earlier, both Deleuze and Derrida have criticized Heidegger's thought of being for its inability to think genuine singularity, an inability that also implies that Heidegger's thought of being cannot think genuine multiplicity. If we cannot think singularity and multiplicity genuinely (that is, in a nonrepresentational way), can we really say that we have overcome *metaphysics*? The final metaphysics, of course, that Heidegger considers is that of Nietzsche, the metaphysics of the will to power, the very metaphysics that supports the contradiction that I just mentioned. Only a concept of life reconceived in terms of *powerlessness* allows us to think singularity and multiplicity. So, it seems to me that, at this moment, for these two reasons—one political, the other philosophical—we must develop something like a neo-vitalism.

Here I shall lay out the *structure* of this neo-vitalism, or what we might call "life-ism."[3] The essay will take place in four steps. First, I will argue that, despite differences,[4] Heidegger's conception of Nietzsche's idea that life is will to power,[5] and Foucault's conception of the modern regime of power as bio-power are similar if not identical conceptions.[6] Both will to power and bio-power are bound up with the Cartesian conception of subjectivity. Coining a term, we can say that "bio-will to power" is the most current and dangerous form of metaphysics, the form that is a mere reversal of Platonism. It will then turn out that Husserl's phenomenological concept of *Erlebnis* is contemporaneous with bio-will to power. So, second, following the well-known distinction between lived-experience (*le vécu*) and the living being (*le vivant*), we will examine the ambiguity in which phenomenological lived-experience consists. To overcome the Cartesian conception of life as subjectivity, it is necessary to disambiguate this ambiguity. The disambiguation occurs by means of a minuscule but invincible hiatus. That a hiatus or a limit is found in life means that death is a teeming presence. Our third step, then, will attempt to follow an opening in Heidegger's thought, in particular, in his 1929 address "What Is Metaphysics?"[7] This address turns death into a process within life; death becomes *Verendlichung* ("finitization"). But our understanding of "finitization" will come mostly from Foucault. Therefore, in a fourth section, we will determine the living being through "mortalism," the "mortalism" that appears in Foucault's two chapters on Bichat in *The Birth of the Clinic*. Inseparable from "life-ism," as we shall see, is a kind of "mortalism." The conclusion, the fourth step, will turn to the fact that "life-ism" follows the line of the inability to preserve and enhance life. Thus it resists the regime of bio-will to power and tries to twist free of metaphysics once and for all.

Life as Will to Power and as Bio-Power (Subjectivism)

It is possible to determine *four similarities* (if not identities) between Heidegger's conception of life as the will to power in Nietzsche (in "Nietzsche's Word 'God is Dead'") and Foucault's conception of bio-power (in *The History of Sexuality, Volume I*). The *first* similarity is the most obvious. Heidegger conceives will to power and Foucault conceives bio-power within the context of the movement of anti-Platonism in the "modern age" (NW 218/63; HS1 155/117).[8] For both Foucault and Heidegger, in the modern age what was above, the second world of forms or the super-sensory, is being pulled down—into life.[9] The collapse of the *super*-sensory into the sensory, into life, is anti-Platonism. The *second* similarity is that both Heidegger and Foucault understand the transformation from Platonism into anti-Platonism as a transformation in the conception of vision.[10] The transformation of Platonism into anti-Platonism means, of course, that, instead of the super-sensory (which no longer expends life), the new highest value is "*super*-abundant life" (NW 226/70, my emphasis). Now, according to Heidegger, super-abundant life, which defines the will to power, is not just the highest value; it itself is an "appraising";[11] it posits values (*Wertsetzung*) (NW 226/70).[12] According to Heidegger, value, in Nietzsche, means "the point in view [*Augenpunkt*] for a seeing that aims at something" (NW 227/71). Heidegger immediately connects this seeing (the point at which the eye aims) to representation (*Vorstellung*). This seeing does not just look and let pass; it sees only insofar as it has seen, meaning that, insofar as it has seen, it has brought the object of vision back and has set (*stellen*) it in front of (*vor*) the seeing, has re-presented it and has posited (*sich vor-gestellt und so gesetzt hat*) the object of vision as such (NW 228/71). Values, then, in Nietzsche according to Heidegger are something posited by means of a re-presentation; there are no values prior to the positing of them. Values are posited by a "gaze," which sets up the point that becomes the "aim in view." According to Heidegger, "Aim, view, field of vision, mean here both the object of vision and the vision, in a sense that is determined out of Greek thought, but that has undergone the change of idea from *eidos* to *perceptio*" (NW 228/72). In other words, what is seen must be an object and an object made *certain* by reckoning and measurement, that is, fixed into what remains as what is "constantly presencing" (*beständig Anwesenden*) (NW 238/83).

Apparently echoing Heidegger or Nietzsche, in *The History of Sexuality, Volume I*, Foucault states that bio-power "distributes the living in the domain of value and utility. Such a power has to qualify, measure, *appraise* [*apprécier*], and hierarchize" (HS1 189/144, my emphasis). In order to appraise, bio-power needs continuous regulatory and corrective mechanisms, mechanisms that are primarily mechanisms of discipline.[13] Moving back, then, to his 1974 *Discipline and*

Punish, we see that Foucault concludes his discussion of discipline with Bentham's conception of the prison as the Panopticon. Just by means of the literal meaning of this word, we can see that power in Foucault, as in Heidegger, concerns vision: optics.[14] For Foucault, Panopticism is a *general way* of ordering bodies, surfaces, lights, and gazes so that power operates automatically and without the intervention of a particular individual like a sovereign (SP 242/207). Appearing in the nineteenth century, Panopticism amounts to a break with antiquity. To speak like Heidegger, we can say that we have the idea no longer as *eidos* but rather as *perceptio*.[15] As Foucault says, the spectacle (temples, theaters, and circuses) solved the *ancient* problem of making it possible for a multitude of people to inspect a small number of objects (SP 252/216). The *modern* problem is the reverse: "to procure for a small number, or even a single individual, the instantaneous *view* [vue] of a great multitude" (SP 252/216). But, since, in the Panopticon, the prisoners are only ever seen and the guard only ever sees, while remaining invisible to the prisoners, Panopticism in fact dissociates the dyad seeing/being seen (SP 235/201–2). Panoptic architecture means that the prisoner is only ever "the object of information" and "never a subject in communication" (SP 234/200). Being only ever an object of vision, each prisoner is "constantly visible" in a state of "permanent visibility"; the prisoner, in other words, is posited. Therefore, "from the viewpoint of the guard," the prisoners are "a multiplicity that can be numbered and controlled" (SP 234/201), appraised. Each prisoner is an object made certain—Foucault calls the Panopticon a "house of certainty" (SP 236/202)—that is, fixed into what remains as "constantly presencing" (*beständig Anwesenden*).

The *third* similarity concerns the aim in view that is posited, the values themselves—in other words, the objective (note the optical sense). According to Heidegger, the objective of life as the will to power is the "preservation and enhancement" of power (*Erhaltung* and *Steigerung*) (NW 229/72). For Nietzsche, according to Heidegger, preservation and enhancement are the two fundamental and inseparable tendencies of life (NW 229/73). Life preserves itself in order to grow and enhance itself, in order to enhance its power. But enhancement is possible only where a "stable reserve" or a "standing reserve" (*Bestand*[16]) is already being preserved as secure. The requirement for the enhancement of a standing reserve of power means, according to Heidegger, that the will to power in Nietzsche must be distinguished from a mere striving or desire for power based on the feeling of a lack. If the will to power were the desire *for* power, one would have willing on one side and power on the other (NW 233 / 76).[17] Instead, the will to power in Nietzsche is "commanding" (*Befehlen*): "Commanding has its essence in the fact that the master who commands has conscious disposal over the possibilities for effective action" (NW 234 / 77). Thus what the will wills it has already; "it super-enhances itself" (*Er übersteigt sich selbst*; NW 234 / 77). The will to power in Nietzsche, according to Heidegger, is power-enhancement; it commands for itself only power and more power, *super*-power (NW 234–35 / 78).

Like Heidegger, who distinguishes a mere striving for power from the will to power, Foucault in *The History of Sexuality, Volume I* distinguishes the juridical power of the sovereign from bio-power. The privileged characteristic of the sovereign's power (going back to ancient times) is the right to decide life and death (HS1 178 / 136). But for Foucault, sovereign right means that this kind of power is exercised by means of "exacting" or "subtracting" the proverbial "pound of flesh" (HS1 178–79 / 136; see also 118 / 89).[18] Juridical power therefore takes away life rather than preserving or enhancing it. But also, because juridical power does not enhance life, the effect of juridical power is obedience (HS1 112 / 85). In this kind of power, one is constrained and forced to submit. Power is taken away, and then there is a desire for power based in a feeling of a lack of power. Following Heidegger's formula, we can say that here, with juridical power, will (or desire) and power are on two opposite sides. Yet in Foucault there is bio-power. The mechanisms of bio-power have the function of organizing and optimizing the forces under the control of those mechanisms. In other words, bio-power is a "power aimed at producing forces, at making them grow, and at ordering them, rather than a power devoted to barring them, to making them compliant, or to destroying them" (HS1 179 / 136). Again borrowing a formula from Heidegger, we can say that bio-power consists in positing conditions of preservation (the maintenance of the biological existence of a population) and enhancement (optimizing and multiplying life). The appropriation of this formula lets us see that what Foucault calls "bio-power" is, as Heidegger would say, a "commanding." In bio-power, there is no desire or will *for* power. Rather, as Foucault says, what is demanded and what serves "as an objective is life, understood as the basic need, man's concrete essence, the accomplishment of his virtualities, a plentitude of the possible" (HS1 191 / 145). Foucault is saying that modern society does not strive; "it super-enhances itself."[19]

The final (fourth) similarity is that both Heidegger and Foucault conceptually link the modern concept of power to Cartesian subject-ivism. For Heidegger, the connection to Descartes occurs by means of value-positing, which must be certain. The certainty then leads to the *ego cogito* as that which presences as fixed and constant, the sub-ject as self-consciousness (NW 238/82–83). Similarly, Foucault claims that "another way of philosophizing" appears due to the expansion of the role of confession in the classical age (from Des-cartes to Kant) and then in the modern age: "seeking the fundamen-tal relation to the true . . . in the self-examination that yields, through a multitude of fleeting impressions, the fundamental certainties of consciousness" (HS1 80/59–60).[20] In particular, according to Fou-cault, Cartesianism appears in the nineteenth century, when psychia-try faced the problem of trying to constitute itself as a science, as "a confessional science"; "the long discussions concerning the possibility of constituting a science of the subject, the validity of introspection, lived-experience [*vécu*] as evidence, or the presence of consciousness to itself were responses to this problem" (HS1 86/64). For Fou-cault, therefore, the modern form of Cartesianism that is involved in bio-power is Husserlian phenomenology and the concept of lived-experience. But the same is true for Heidegger. In "Nietzsche's Word 'God is Dead,'" Heidegger implies that nihilism ends in phe-nomenology (NW 209/54).[21] Phenomenology's unthought and in-vincible presupposition, Heidegger says, is the "un-positing of the supersensible," which in turn *reduces* the difference between the sen-sible and the supersensible.

Before we leave this section, let me summarize the four similarities between Heidegger's interpretation of Nietzsche's will to power and Foucault's bio-power. First, both conceptions occur in the modern epoch, which is the epoch of anti-Platonism. Second, both concep-tions, being modern, imply a transformation of vision into positing and constant presence. Third, bio-power and will to power are com-manding, meaning that the will in each conception super-enhances the power that it already has; bio-will to power is the will to more and more power (*super*-abundant life). Finally, fourth, both Heideg-ger and Foucault associate the phenomenological concept of *Erlebnis* with bio-will to power. So let us now turn to the concept of lived-experience.

Life as Lived-Experience (Immanence)

Expanding on Descartes's idea of methodical doubt, Husserl, of course, invents the phenomenological *reduction*. What defines the phenomenological concept of lived-experience is an "ambiguity" between *noesis* and *noema*, between what is *reell* and what is *irreel*, between what is intentional and non-intentional, to use the terminology of *Ideas I*.[22] We know that Husserl understands lived-experience in terms of temporalization, absolute temporalization being a form of auto-affection. We also know that Heidegger, in his 1929 *Kant and the Problem of Metaphysics*, understands temporalization as auto-affection, but within the context of the question of finitude.[23] So, what I have just described very quickly is a movement from lived-experience as an *ambiguity* between subject and object (noesis and noema). But the ambiguity results from lived-experience being temporal auto-affection. And this auto-affection, being temporal (being human and not divine, as in Aristotle's thought thinking itself), must be finite. How are we to understand finitude?

The question of finitude is precisely Foucault's question in *Words and Things*, chapter 9, "Man and His Doubles." Here Foucault contrasts finitude in the modern epoch to finitude in the classical epoch. In the classical epoch (again from Descartes to Kant), finitude is conceived in terms of the infinite; there was a *metaphysics* of the infinite, of God. In the modern epoch, however, according to Foucault, finitude is conceived in terms of itself. Now we have, as Foucault says, "the end of metaphysics." The modern epoch is indeed the epoch of anti-Platonism. In the modern epoch, we have *man* (instead of God). According to Foucault, man is finite in two ways. As an object of knowledge (say, in the human sciences), man is finite insofar as he is subjected to life, language, and work. But as a subject of knowledge man is also finite, since the forms in which he knows life, language, and work are finite. Again, man is finite in two ways, but the two are the *same*. As is well known, the sameness of finitude means that "man" is in the middle of a series of doubles: the foundation and the founded; the empirical and the transcendental; the thought and the unthought; and the return and retreat of the origin. The doubling (as in a mirror image) implies that we can say, according to Foucault, that I *am* this life since I sense it deep within me, but also I can say that I am *not* it since it envelops me and grows toward the imminent moment of death (MC 335 / 324–25). In "man and his doubles" then, we have an auto-affection that is the same (MC 326 / 315) but that is also other, also hetero-affection. As with the phenomenological concept of lived-experience, the relation—the kind of "synthesis" that "man" represents—is ambiguous (MC 332/321). Yet Foucault claims that the sameness of the doubles, indicated by the conjunction *and* (*et*[24]), is possible only on the basis of "un écart infime, mais invincible," "a minuscule, but invisible hiatus" (MC 351/340). For Foucault, this *écart* differentiates the senses of the ambiguity; it disambiguates the auto-affection of man;[25] it disambiguates the immanence of lived-experience.[26]

In *Words and Things*, however, Foucault provides no argumentation for the claim that *un écart* establishes difference. We can find the argumentation, I think, *only* in Derrida's *Voice and Phenomenon*, his 1967 study of Husserl. We have seen this argumentation before, but it is so important that it must be repeated. Only in this context can we see its full impact. Derrida argues that, when Husserl describes lived-experience, even absolute subjectivity, he is speaking of an interior monologue, auto-affection as hearing-oneself-speak. According to Derrida, hearing-oneself-speak is, for Husserl, "an absolutely unique kind of auto-affection" (VP 88/78), because there seems to be no external detour from the hearing to the speaking; in hearing-oneself-speak there is self-proximity. I hear myself speak immediately in the very *moment* that I am speaking. According to Derrida, Husserl's own description of temporalization undermines the idea that I hear myself speak immediately. Husserl describes what he calls the "living present," the present that I am experiencing right now, as being perception, and yet Husserl also says that the living present is thick. The living present is thick because it includes phases other than the now—in particular, what Husserl calls "retention," the memory of the recent past. Yet retention in Husserl has a strange status, since Husserl wants to include it in the present as a kind of perception, and at the same time he recognizes that it is different from the present, being a kind of nonperception. For Derrida, Husserl's descriptions imply that the living present, by always folding the recent past back into itself, involves a *difference* in its very midst (VP 77/69).[27] In other words, in the very moment when silently I speak to myself there must be a minuscule hiatus differentiating me into hearer and speaker. There must be *un écart* that differentiates me from myself, *un écart* without which I would not be a hearer *as well as* a speaker. And this *écart* is in the very moment of hearing myself speak. Derrida, of course, stresses that "moment" or "instant" translates the German *Augenblick*, which literally means "blink of the eye." When Derrida stresses this literal meaning, he is in effect "deconstructing" auditory (and tactile) auto-affection into visual auto-affection. When I look in the mirror, for example, it is necessary that ("il faut que," Derrida says) I am "distanced" or "spaced" from the mirror. I must be distanced from myself so that I am able to be *both* seer *and* seen. The *space* between, however, remains "obstinately invisible." Remaining invisible, the space gouges out the eye, blinds it. What Derrida is trying to demonstrate here is that this "spacing" or blindness is essentially necessary for all forms of auto-affection.[28]

The essential spacing (*espacement*) that Derrida demonstrates brings forward several important consequences for how we conceive the immanence of lived experience. Spacing in fact transforms immanence itself. It frees immanence, which is life itself, from being immanent to consciousness. Both subject and object (both consciousness and matter) are now constituted by spacing, which itself is neither subjective nor objective. Immanence, nevertheless, is not ambiguous; the subjective side and the objective side are differentiated by the minuscule hiatus, by the blind spot. But we can go one step farther. Being a blind spot in the middle of life, the minuscule hiatus is a kind of dead zone (*la place du mort*). Spacing therefore redefines life as a process of dying.

Heidegger can help us start to understand this process of dying. In the lecture course from the academic year 1929–30, *The Fundamental Concepts of Metaphysics*, Heidegger attempts to awaken in us (GA29/30, 103 / 69) the mood (*Stimmung*) of boredom, "profound boredom." Because profound boredom is boredom with all things, with the whole, with the whole world, Heidegger turns to the question of world. Famously, in order to answer the question of what is world, Heidegger engages in a comparative examination of the relation of the world to the stone, the animal, and man. The comparison depends upon understanding the essence of the living being. At the end of the discussion of the animal, a discussion that covers almost one hundred pages, Heidegger concludes that this essence is still a problem and will remain so "until and unless we also take into account the fundamental phenomenon of the life process [*Lebensprozesses*] and thus death as well" (GA29/30, 396 / 273).[29] This "life process and death as well" is what Heidegger in "What Is Metaphysics?" calls *Verendlichung* ("finitization").

Verendlichung ("Finitization")

In the inaugural address Heidegger answers the question "What is metaphysics?" by asking a specific metaphysical question: "How is it with the nothing?" He comes to this specific metaphysical question by means of a reflection on the sciences; science is concerned with beings and nothing else besides. Thus the sciences seem to depend on this "nothing else besides" for their self-understanding. They depend, in other words, on this precise metaphysical inquiry into the nothing; they depend, in short, on metaphysics. Concerning the relation between metaphysics and science, Heidegger opens the 1949 introduction by describing the old idea of a tree of knowledge, with the tree being science and the roots being metaphysics. He starts the introduction in this way because, in his address, Heidegger says that "the rootedness of the sciences in their essential ground has withered [*abgestorben*]" (GA9, 104/83). Thus the address is asking about metaphysics, the soil or ground of the sciences, in order to revitalize the roots of the sciences. But more importantly, in the address Heidegger claims that science determines our existence (here the word is, of course, *Dasein*). The question is: "What is happening to *us*, essentially, in the grounds of our existence, when science has become our passion?" (GA9, 103/82, my emphasis). The entire address revolves around this "us." The revitalization of the roots of science, that is, the revitalization of metaphysics, depends therefore upon the "completion" of the "transformation of the human being into . . . Dasein" (GA9, 113/89; Heidegger's hyphenation of *Dasein*). This transformation means the transformation of *the subject* into the *Da* of *Dasein*, that is, into the place of the nothing (GA9, 113Na, 373Na).

In order to complete the transformation of "us" into *Da-sein*, and thereby to revitalize the roots of science, we must pursue a response to the question of how it is with the nothing. As always with Heidegger, the inquiry starts with our everyday understanding of the nothing, with our everyday definition: "The nothing is the complete negation [*Verneinung*] of the totality [*Allheit*] of beings" (GA9, 108/86). Importantly, immediately after producing this everyday definition, Heidegger says, "Does not this characterization of the nothing ultimately provide an indication [*Fingerzeig*] of the direction [*Richtung*] from which alone the nothing can come to meet us?" As we will see, our encounter with the nothing happens by means of a pointing finger, *ein Fingerzeig*. And this everyday definition indeed gives us a sign in the right direction. On the one hand, it implies that we must have access to the "totality of beings" or to "beings as a whole," and, on the other hand, somehow this totality or whole is negated. We imagine the whole of beings in an idea, and then we "think" it negated (GA9, 109/86–87). Heidegger calls this two-sided definition "the formal concept of the imagined nothing." But this formal concept is not the nothing itself.

Thus we are still confronted with the problem of access to the nothing itself, especially in light of the fact that we are "finite essences" (*endliche Wesen*) (GA9, 109/86). But obviously insofar as we exist, we find ourselves all the time in the midst of beings as a whole. This disposition of being in the midst of beings as a whole (in the midst of the world, in other words) all the time becomes particularly evident in "profound boredom." In his address Heidegger turns to moods (*Stimmungen*). Now, in the 1929–30 lecture course, Heidegger distinguishes "profound boredom" from more superficial kinds. But the central difference is that profound boredom is not boredom with one thing—a theater performance, for example—but with all things. Profound boredom is indeterminate and makes one be indifferent (*gleichgültig*) in regard to everything, in regard to beings as a whole. But in light of this indifference, as Heidegger says in "What Is Metaphysics?" profound "boredom manifests beings as a whole" (GA9, 110/8).[30] Therefore, if we recall the "pointer," *der Fingerzeig*, that the everyday definition of the nothing provided, profound boredom has given us the totality of beings or beings as a whole. Following this pointer, we would then say that, in order to encounter the nothing, what remains is some sort of negation of beings as whole. But this is not what Heidegger says; instead, he says, "we will now come to share even less in the opinion that the negation of beings as a whole that are manifest to us in the mood places us before the nothing" (GA9, 110/87–88). There will be no negation of beings as a whole. We will have a mood that manifests the nothing *and* does not negate beings as a whole. This mood, which Heidegger calls "fundamental" (*Grundstimmung*), is, of course, anxiety (*Angst*).

What is striking about the descriptions of anxiety that Heidegger gives in "What Is Metaphysics?"—these descriptions extend over seven pages (GA9, 111–18/87–93)—is their difference from the ones given in *Being and Time*. The most obvious difference is the fact that nowhere in "What Is Metaphysics?" does Heidegger speak of death.[31] We must keep this point in mind—no mention of death—if we want to understand the opening to a different kind of thinking that we can find in this text. As in *Being and Time*, here in "What Is Metaphysics?" Heidegger distinguishes fundamental anxiety from fear. While fear is always felt in the face of a determinate something, fundamental anxiety is felt in the face of no determinate thing. Thus fundamental anxiety is intimately connected to profound boredom. Both moods, fundamental and profound, consist in indeterminateness. But in "What Is Metaphysics?" Heidegger insists on the indeterminateness of that in the face of which one feels anxious, saying that it is essentially impossible to determine it. He recalls the familiar German expression "Ist es einem unheimlich," which is rendered in English as "One feels uncanny." In the contemporaneous lecture course, Heidegger focuses on a similar phrase for profound boredom, "Es ist einem langweilig" (GA29/30, 202 / 134). For *both* profound boredom and fundamental anxiety, we would have to admit that "we *cannot* say what it is before which one feels uncanny" (GA9 111 / 88, my emphasis) and that we *cannot* say what it is before which one feels bored. This *nicht können*, this "cannot," brings us to the essential feature of fundamental anxiety; fundamental anxiety is essentially an experience of impotence (*Ohnmacht*) (GA9, 113 / 90). Therefore, what Heidegger says in the lecture course about profound boredom holds for fundamental anxiety: "The 'it is boring for one' has already transposed us into a realm of power [*einen Machtbereich*] over which the singular person, the public individual subject, no longer has any power" (GA29/30, 205–6 / 136). To say this again, the essential feature of fundamental anxiety is powerlessness. This powerlessness is how the nothing becomes manifest (*offenbar*).[32]

Indeed how, precisely, does the nothing manifest itself? Heidegger says that the nothing becomes manifest "at one with" (*in eins mit*) beings as a whole (GA9, 114 / 89–90). With regard to this "one-ness" of the manifestation of beings as a whole "with" the nothing, it is necessary to recall why metaphysics must be put in question: in metaphysics, there is a "persistent confusion of being and beings." Therefore, the *in eins mit* must not be a confusion of this sort; it must not be a mixture of being and beings; there must be no ambiguity here.[33] Being and beings must be differentiated. But for Heidegger, this differentiation does not involve negation; the manifestation of the nothing is not an annihilation of beings as a whole. Instead, we have Heidegger's famous (or infamous) statement "Das Nicht selbst nichtet," which is rendered in English as "the nothing itself nothings." It is important to see that this statement is positive, involving no negative adverb. But now our question of how the nothing manifests itself has become a question of how the nothing nothings.

The nothing nothings by being "essentially repelling" (GA9, 114 / 90). The English word *repelling* renders the German *abweisend*, and Heidegger also speaks of the *Abweisung* of the nothing, which the English translation renders as the "repulsion" of the nothing. But the "repulsion" of the nothing, its *Abweisung*, is also, according to Heidegger, a *Verweisen*, which the English translation renders as "gesture." Clearly, the two terms *Abweisung* and *Verweisen* are connected, and Heidegger indicates their connection in a note he added to the 1949 edition of "What Is Metaphysics?" (GA9, 114Na/90Nb). The nothing nothings, therefore, by means of what we would have to call *Weisung*, pointing with the finger or leading by the hand, a directive, *eine Weisung*.[34] Or perhaps we would have to say that the nothing happens by means of "ference," "ference" being a way of rendering in English (or in French) the sense of *Weisung*.[35] *Ference* means to carry or bring, as a hand might direct and bring. But for the hand to point and direct, it must not be holding anything. We can now understand more fully the powerlessness in which anxiety consists. The "repulsion," the *Abweisung* of the nothing is a "de-ference," an *Ab-weisung*, a pointing away.[36] It makes the one undergoing anxiety *defer*; there is "a shrinking back before [*ein Zurückweichen vor*]." When one shrinks back before, the hand of the one experiencing anxiety must let go and no longer grasp (*erfasst*). Indeed, Heidegger tells us that in anxiety there "is no kind of grasping [*Erfassen*] of the nothing" (GA9, 113/89). Beings as a whole then slip away out of one's hands: "no hold remains [*es bleibt kein Halt*]" (GA9, 112/99). Yet at the same time there is a gesture, a *Ver-weisung*, *to* beings as a whole; beings as a whole are directed to sink away into indifference;[37] they are referred to themselves. Referring to themselves, beings no longer point away from themselves to something else; they are no longer present as referring away. This self-reference means that, in the *Weisung* of the nothing, we are involved in a different system of *Verweisung* than the system described in paragraph 17 of *Being and Time*, called "Verweisung und Zeichen." In the *Weisung* of the nothing, one is "directed [*verweist*] precisely to beings" (GA9, 116/92), but beings are "frail" (*hinfällig*). This frailty means that beings are *not as* "for the sake of which [*Umwillen*]"; instead, they are pointed out *as such*. By gesturing, by the pointing finger of the hand that no longer grasps (a hand from which we cannot separate a kind of seeing or story of the eyes), beings as whole, so to speak, "presence," and when they presence, they are manifest "in their full but heretofore concealed strangeness as radically other [*Befremdlichkeit als das schlechthin Andere*]" (GA9, 114/90). What has happened here is that the pointing finger of the hand that no longer grasps partitions beings off from their "for the sake of which" (from their mundane meaning, in other words; cf. GA9, 53/122), in order that they may manifest themselves as such, as being as a whole. In the 1929–30 lecture course, Heidegger tells us that the world is the manifestation of beings as such as a whole (GA29/30, 411/284, 512/353). And, in the 1928 lecture course on Leibniz's logic, we learn that the world is a nothing, indeed, the *nihil originarium*.[38] Through the nothing that is the world, one sees that "they are beings—and not nothing" (GA9, 114/90). By means of the *Weisung* of the nothing, the ontological difference has been established.

How *does* Heidegger conceive the ontological difference? Gadamer reports that in the fourth edition of the postscript to "What Is Metaphysics?" Heidegger wrote, in *Pathmarks*, "It belongs to the truth of Being that Being certainly presences [*west*] without beings, but there is never a being without Being." According to Gadamer, at this point Heidegger was seduced by the Platonic idea of *Chorismos*, separation, and thus by a certain kind of dualism.[39] Then, in the fifth edition, Heidegger changed this sentence to read: "Being never presences without beings; there is never a being without Being."[40] This re-wording is more consistent with Heidegger's numerous reflections on the phrase "the being of beings." These reflections always indicate that Heidegger conceives the ontological difference neither as an identity nor as a duality.[41] We shall return, of course, to this important point concerning the ontological difference when we turn to Foucault. But no matter how it is conceived, the ontological difference is established because the nothing itself forces *us* to shrink back and to release beings. We must now concern ourselves with this "us."

Since we who are experiencing anxiety are also beings, we too slip away. We too become strange or foreign, with our self or person or individuality slipping away out of our hands. This slipping away of us is why Heidegger says that fundamental anxiety and profound boredom occur to "one," *einem*. In "What Is Metaphysics?" Heidegger says, "At bottom [*im Grunde*] therefore it is not 'you' or 'me' who are uncanny, rather it is this way for 'one'" (GA9, 112/89). This *einem* means that these moods occur anonymously and yet singularly: "one," *einem*. When we become "one," or better, "ones," then, as Heidegger says, only "pure *Da-sein* . . . is still there [*da*]." The "there" or place of existence is being held out into the nothing (GA9, 115/91). And that place also means being held out *over* the nothing or, more simply, over nothing, over the abyss (the *Abgrund*). This groundless ground, which is *Da-sein*, turns us into, as Heidegger says, a "lieutenant" in the literal sense of being a placeholder, *ein Platzhalter*, for the nothing (GA9, 118/93). Perhaps this idea of the lieutenant is the basis for Heidegger's later reflections on dwelling (*wohnen*).

Yet the idea that we become a placeholder for the nothing by being singularized brings us to the quote that, I believe, is the heart of this address: "We are so finite that we cannot even bring ourselves originally before the nothing through our own determination and will. So abyssally [*abgründig*] does finitization [*Verendlichung*] entrench itself that our most proper and deepest limitation [*Endlichkeit*] refuses to yield to our freedom." *Verendlichung* is the opening beyond metaphysics, even beyond Heidegger's thought of being. *Verendlichung is not*—it is not beings, being, or their negation. The suffix of the word, *Verendlich-ung*, suggests that it is a process, a process of finitude; the prefix of the word, *Ver-endlichung*, suggests that it is a becoming-finite, a becoming-finite, moreover, that is repeated indefinitely, a "re-finitization."[42] The place that we are is an indefinite becoming-finite, and this fact explains why Heidegger never mentions death in "What Is Metaphysics?" Unlike what we see in *Being and Time* (division 2, chapter 1), in "What Is Metaphysics?" death is no longer a possibility in relation to which one can be an "anxious freedom toward" and thereby grasp one's ownmost existential possibilities. Our deepest limitation refuses to yield to our freedom.[43] Heidegger indeed seems to have developed an idea inspired by this saying: "As soon as a human being is born, he is old enough to die."[44] In *Being and Time*, of course, Heidegger makes a differentiation within death.[45] The English translation of *Being and Time* uses "perishing" in order to render *verenden*: "we call the ending of what is alive *perishing*" (Heidegger's italics: "Das Ended von Lebenden nannten wir *verenden*"). Heidegger further says that *Dasein*, that is, human existence, perishes too, since it is physiological, like all living beings. But *Dasein*, according to Heidegger in *Being and Time*, does not simply perish. Two things can be said about this differentiation. On the one hand, the verbal use of *verenden* implies a generalized death, which in "What Is Metaphysics?" (two years after *Being and Time*) becomes a process: *Verendlichung*. Within this generalized process it is possible to make differentiations among living beings or animals. On the other hand, in *Being and Time*, *Dasein*'s "properly dying" (*eigentlich sterben*) seems to move toward a "constant thinking about death."[46]

Heidegger immediately says that this "brooding about death" is not what he means with "properly dying." But the possibility of impossibility that is death must be "understood as such," revealed as such. *Verendlichung*, however, is precisely something that cannot be understood, since it is "something that possesses us" (cf. GA29/30, 427–28 / 294–95).[47] *Verenðlichung* implies, first, that death is not an absolute limit; rather, as becoming-finite, death is a relative limit. The limit has been distributed throughout existence. Second, death is not the end—in *Being anð Time*, Heidegger seems to continue to conceive death as an end, "the end of *Dasein*"—rather, as becoming-finite, death has become multiple, has been multiplied indefinitely. We have become the place where ends, where "de-limitation," *Ver-enðlichung* happens indefinitely. Third, death is not a possibility of no longer existing but rather, as becoming-finite, death is actually life itself.[48] Indeed, with this *Verenðlichung*, we have reached the place of the subrepresentational and the in-formal. In his book on Foucault, Deleuze creates a phrase that we can apply to this place outside of representation and form: "Bichat's zone."[49]

Life as the Living Being (Mortalism)

As Deleuze says, "from *The Birth of the Clinic* on Foucault admired Bichat for having invented a new vitalism."[50] Indeed, in *The Birth of the Clinic*, Foucault devotes two chapters to the innovations that Bichat brought about in relation to the concept of life at the *threshold* between the classical epoch and the modern. According to Foucault, we reach modern biology when the classifiable characteristics of the living come to be "based upon a principle alien to the domain of the visible" (MC 239 / 227). While the classifications involved in classical natural history related visible form to visible form—for example, the visible form of a bird to the visible characteristic of wings— modern biology relates visible form to the functions essential to the living, functions that are themselves buried deeply within the body of the living. These functions form a "hidden architecture" of the *in*visible (MC 242 / 229). The discontinuity, which consists in a break between a relation of visible to visible and a relation of visible to invisible, brought forth another, a discontinuity in medical perception. The gaze becomes, as Foucault says, "the great white eye of death" (NC 147 / 144). Bichat had, of course, become famous during his own time and later because he had "opened up a few corpses."[51] Literally, "autopsy," "auto-opsis," is a "seeing with one's own eyes," not one's own life, but the death of the living. The autopsy is not a lived-experience (NC 175 / 170) but rather a dead-experience.

Now, Foucault points out that during this period—the Enlightenment—autopsies were possible *immeðiately* after the moment of death. Consequently, the last stage of the process of the disease, the last stage of "pathological time," nearly coincided with the first stage of death, the first stage of "cadaveric time" (NC 143 / 141). At this moment, decomposition had only just started. Because the effects of decomposition were nearly suppressed, death became "a landmark without thickness," "the vertical and absolutely thin line that separates but allows the series of symptoms to be related to the series of lesions" (NC 143 / 141). Bichat realized that at the moment of death one is able to observe the final symptom in a series of symptoms—the final point in the series is still visible—*and*, in the autopsy opening the body, one is also able to observe the series of lesions, a series that, prior to death, had been hidden in the patient's body, invisible. In other words, by means of the immediate autopsy, Bichat was able to look at the "principle alien to the domain of the visible." Bichat made what was invisible when alive visible through the autopsy, through *death*. This is, for Foucault, Bichat's *first great innovation*: the perception of death allowed Bichat to give a more rigorous and therefore instrumental definition of death (NC 143 / 141, 149 / 146). He made the medical gaze pivot away from elimination of disease, cure, and preservation of life, toward death, demanding of death that it give an account of life and disease (NC 148–49 / 146).

Because of having done autopsies, Bichat was also able to distinguish two different but inter-related series. If one opens up the body at the very moment of death, one is able to see the series of the disease, its progress across tissues; this progress is the "morbid process." But there is also a series of phenomena that announce the coming of death: for instance, muscular flaccidity, Foucault tells us, would not take place without a disease. But the flaccidity is not the disease itself; it accompanies any chronic disease. Therefore, the increase in muscular flaccidity "doubles" the duration of the disease with an evolution that indicates, not the disease, but the proximity of death (NC 144 / 141). Foucault calls a process like the increasing flaccidity of muscles "mortification." Mortification runs beneath the morbid processes with which it is associated. According to Foucault, however, the signs of mortification are different from the symptoms of the disease; symptoms allow one to predict the outcome of the disease toward health or illness. The signs of mortification, however, simply show a process in the course of accomplishment; they are phenomena, as Foucault says, of "partial or progressive death": "long after the death of the individual, minuscule, partial deaths continue to dissociate the islets of life that still subsist" (NC 144–45 / 142).

For Foucault, this idea of a "moving death" is Bichat's *second great innovation* in regard to the concept of life. These processes indicate the *permeability* of life by death. Foucault says, "Death is therefore multiple, and dispersed in time: it is not the absolute, privileged point at which time stops and moves back; like disease itself, death has a teeming presence" (NC 144 / 142). The idea of partial deaths leads to Bichat's third innovation.

Thanks to the great white eye of death—the autopsy—disease has, Foucault says, "a mappable land" or "place" (NC 151 / 149). The living is defined by the *spacing* of disease as a "great organic vegetation" with "a nervure," with "its own forms of sprouting, its own way of taking root, and its privileged regions of growth" (NC 155 / 152–53). Spatialized in this way, pathological phenomena take on the appearance of living processes. That disease is a living process means that it is inseparable from life; it is no longer an event or nature imported from the exterior of life. Disease is an internal deviation of life (NC 155 / 153). But, being an internal deviation of life, disease is also organized according to the model of a living individual; there is, for example, a life of cancer. Both of these consequences—that disease is a living process and that it is a life—for Foucault indicate that disease is now understood as pathological *life* (NC 156 / 153). He says, "From Bichat onwards, the pathological phenomenon was perceived against a background of *life*, thus finding itself linked to the concrete and obligatory forms that life takes in an organic individuality" (NC 156 / 153, Foucault's emphasis). For Foucault, with the idea that life is the ground of disease, we have Bichat's *third innovation*, his definition of life as the set of functions that resist death. For Bichat, life "is not a set of characteristics that are distinguished from the inorganic, but the background against which the opposition between the organism and the non-living may be perceived, situated, and laden with all the positive values of conflict" (NC 157 / 154). Defining life as a conflict with the nonliving means that life, being a process of degeneration, is at the limit auto-destruction, and not preservative (NC 160–61 / 157); the degeneration of life always moves toward death. Wear and tear (*l'usure*), Foucault says, "is the form of degeneration that accompanies life, and throughout its entire duration, defines its confrontation with death" (NC 161 / 158). Therefore, for Foucault— this claim is really Bichat's third innovation—death is co-extensive with life. The co-extensivity of death with life is why Foucault says that "vitalism appears against the background of 'mortalism'" (NC 148 / 145).[52]

But how are we to understand mortalism?[53] Foucault says: "al-though . . . Bichat plugged the pathological phenomenon into the physiological process and although he makes the pathological phenomenon derive from the physiological process, this deviation, in the hiatus that it constitutes, and which announces the morbid fact, is founded on death. Deviation in life is of the order of life, but of a life that goes toward death" (NC 158 / 156). Death, Foucault says here, constitutes *un écart*, even *un écart infime*, "a minuscule hiatus" (cf. MC 351 / 340). Always involving the *écart* of death, life always contains disease as a potentiality.[54] Foucault stresses that it is not the case that disease is the source of death; rather, death, in life, has always already begun; there is always already a process of mortification, in which diseases are virtual. The reversal of what we normally think about the relation between life, disease, and death defines mortalism; mortalism is the virtuality, in life, of dying (finitization). The virtuality of dying, however, means that life, or better, "a life," is always potentially a multiplicity of diseases, which, as we have already seen, are themselves modeled on a living individual, on a life: there is a life of cancer. Multiplicity, being the ground of many lives of diseases, contains virtually singularities (NC 159 / 156).[55]

The virtuality of singularities brings us back to the gaze of the autopsy. Bichat's first innovation was the transformation of death into an instrument by means of which lesions that were invisible during life become visible. Here we have what Foucault calls a "visible invisible," in other words, "an invisible that is potentially visible."[56] Yet Foucault calls life "an opaque ground" (NC 170 / 166), meaning that this seeing with one's own eyes reaches a limit. Individual diseases exist only because they unfold in the form of individuality (NC 173 / 168–69). Therefore, in order to describe a life of a disease, medical experience would have to see all the singularities that support it. To determine the singularity of a disease, we would have to see more qualitatively, more concretely, in a way that is more than individualized; there would have to be a greater refinement of vision; we would be going farther and farther into the gaps between the singular points. Yet if these singular points that we are approaching are really singular, our eyes would not be able to recognize them with any already acquired general representations; simply, these representations would be general and not singular. These forms would not work, and there could be no understanding—there could be no positing. A life would be both sub-representational and informal.[57] Through Bichat, Foucault is able to conceive life as a zone in which there is the constant murmur of conflict, but in which the gaze is uncertain of what it is seeing. Here, in "Bichat's zone," we have not a visible invisible but an invisible visible, a visible that remains obstinately invisible.[58] The autopsy, then, is unable to see clearly in the grayness of this zone, and yet it sees more, a strange form of blindness.

Conclusion: Powerlessness and Power

To conclude, let me summarize the structure of life-ism that this chapter has just laid out. Under the pressure of Heidegger's experience of the forgetfulness of being in Western metaphysics, continental philosophy revolves around the question of overcoming metaphysics. Thus continental philosophy (understood as a project) always refers itself back to Nietzsche's idea of anti-Platonism.[59] To overcome Platonism, it is necessary to collapse the *division* (as in the divided line) between this world and the second world of forms. The collapse of the difference not only reduces the forms to life, that is, to immanence, but also shifts the level of the concept of life itself. Prior to the nineteenth century, the spontaneity of life is opposed to the determination of machines, and this opposition between vitalism and mechanism takes place against the background of the fundamental concept of *nature*.[60] With the project, however, of the overcoming of Platonism, the concept of life itself becomes the background or ground for all other oppositions. This shift to the level of ground implies that the traditional (pre-nineteenth-century) problems associated with the concept of life are pushed to the side, problems such as the unity of life (vegetative versus cognitive), the specificity of life (organic versus inorganic), the opposition between finalism and mechanism, the conceptions of evolution. Instead, replacing being as well as nature, life becomes *ultra-transcendental*.[61] The ultra-transcendental concept of life is auto-affection. What distinguishes the twentieth-century concept of auto-affection from all previous ones is that now auto-affection takes place across a limit, across a difference,[62] across a minuscule but invincible hiatus, across a spacing. Due to the limit, auto-affection is finite; or, since life as auto-affection is fundamental, life is "originary finitude."[63] To put this idea another way, below "life-ism" is "mortalism" (NC 148 / 145). The limit in the middle of auto-affection is death. Yet, death is not an absolute limit opposed to life; it is not an end. Rather, death is mobile, a "teeming presence": life, then, is "finitization."

Finitization means that the power of life rests on a powerlessness.[64] The teeming presence of death means that, within the sensing-sensed relation, there is always the nonsensible. If the living is defined by self-relation, then wherever there is *this* relation, there must be a *gap* between the active and passive poles of the relation in order that there might be *two* poles at all. The "auto," then, is always, necessarily, out of joint. The necessary being out of joint means that, in auto-affection, understood as seeing-seen, there is always a kind of fundamental blindness; the eye closes (VP 73 / 65; see also MC 337 / 326).[65] This fundamental blindness, an inability to see, places *powerlessness* right in the very midst of the viewpoint of *power* understood as preservation-enhancement. There is a fundamental a-perspectivism within the perspectivism of the "bio-will to power." Indeed, life is so finite that it cannot overcome this powerlessness through determination and the will. It seems to me that the argumentation that life as auto-affection is always "out of joint" amounts to the only way to start to understand Foucault's concept of points of resistance against bio-power (HS1 126 / 95–96).[66] As Foucault says, "the forces that resisted [bio-power] relied for support on the very thing [bio-power] invested, that is, on life and man insofar as he is living [*vivant*]" (VS 190 / 144). Only where there is blindness (and blindness always makes one uncertain), only where one (a life) is no longer visible to the general Panopticon, is resistance possible. With blindness, there can be no constant presence; there is always invisibility and thus freedom. In other words, from the points of impossibility (not the plenitude of the possible) comes the possible. Only by *following* the line of this powerlessness will we be able to twist free of the most current and dangerous form of metaphysics, the form that is the *mere* reversal of Platonism. Why most dangerous? In the regime of bio-will to power, there is "radical killing" (NW 263 / 108), and "wars have never been as bloody" (HS1 179 / 136).

Conclusion
The Followers

Thanks to the signs, we see more distinctly—and distinctness does not necessarily exclude obscurity—that we live in an epoch of bio-power. The will to the preservation and enhancement of life dominates in the West because the second world of ideas, the Platonic "sun," has set. The division, as in the divided line, has collapsed. Anti-Platonism, which is the negation of the difference between the two worlds, is the most dangerous form of metaphysics. It is the reduction of everything to a kind of "actualism," or, to use a term popular in analytic philosophy today, to a kind of "naturalism." Greek metaphysics, however, is only one part of Western thinking. Contemporary naturalism (the knowledge willed in relation to bio-power) consists, as well, in a negation of Christianity. Through its drive to be reductionistic, naturalism is not only anti-Platonic, but also anti-Christian. Christianity subsumes both Greek metaphysical-religious thought and Jewish metaphysical-religious thought. As early as "Violence and Metaphysics," Derrida speaks of "jewgreek," and there he anticipates the idea of a "deconstruction of Christianity." It is possible that Foucault's four volumes of *The History of Sexuality* count as a deconstruction of Christianity. We must not forget that deconstruction is always a form of critique, even a form of enlightenment. Deconstruction would criticize both the mere reversal of Christianity (the reversal into naturalism) and Christianity itself. Doing so, it would disclose an origin, principle, or *archē* that is not

Christian, or pre-Christian. We do not know in what way the origin would be named; perhaps the name would be Greek, Jewish, perhaps even Islamic. But a deconstruction of Christianity would have to focus on Christianity's catholicism (Romans 2:29). Not being restricted to ethnicity, not being restricted to a part, it crosses borders toward the whole (*kata-holos*); in fact, maybe Christianity concerns nothing but border-crossings. In the event of the Christ, God crosses the border of the human. Crossing this border, God is not reduced to man, to the actual; the incarnation is not a mere negation of God; it is not pure negativism. Here, we would need to open a discourse of "theiology" (a discourse of the divine) and not theology (a discourse of God). Perhaps a "theiology" is what Merleau-Ponty had in mind when he called for a contemporary translation of "large rationalism." The small operations of today's reason, with its calculations of the preservation and enhancement of power, must be put back in relation to something larger than itself. And what is larger is not purely alive, nor is it purely dead. Here we could call upon Valéry's image of the park in his poem "La Jeune Parque": the trees with their "constellations" of leaves and their "constellations" of roots bring the heavens of life into relation with underground rivers of death.[1]

The other side of the deconstruction of Christianity would then be what I have called "life-ism." Life-ism would not be a return to earlier versions of vitalism. Foucault showed in both *Words and Things* and *The Birth of the Clinic* that, at the beginning of the nineteenth century, Bichat's name is the index of the rupture that opened the modern epoch. When Bichat developed his definition of life, he placed life at a deeper level; he replaced nature with life as the ontological foundation. Life itself becomes the background or ground for all other oppositions. This shift to the level of ground implies that the traditional problems associated with the concept of life are pushed to the side, problems such as the unity of life (vegetative versus cognitive), the specificity of life (organic versus inorganic), the opposition between finalism and mechanism, the conceptions of evolution. Instead, replacing being as well as nature, life becomes *ultra-transcendental*. Vitalism, therefore, is an idea that does not belong to our present. We can assemble the characteristics that define the new concept of life. It would not be biological in a strictly material sense; it is not natural life (*zoōn*). Instead, this life, the living, is spiritual. To call life spirit (as opposed to matter) implies conceptualization, information, the virtual, memory. Memory means that language is in life. At the center of the new conception of life, we must place the question of language. Like names and statements, the singularities of a life, of a disease, of a proper name, are repeatable, universalizable. Or, to be more precise, what defines the living is the possibility of forgetting and dying: the discontinuity in repetition. The space opened up by discontinuity refers us to the zone of which we have spoken continuously: the zone of *un écart infime*. But this divergence produced by the iterability of the event—one life—implies the possibility of the monster, the anomaly, error. What defines life is the capacity for error. Therefore life is necessarily capable of becoming animalistic, capable of becoming rogues (*voyous* or *canailles*). This consequence even implies that language itself consists of animals.

But why this word *life*? Unlike the Latin word *vita*, and unlike both the Greek words *bios* and *zoōn*, this Anglo-Saxon word *life* is connected to the German *Leib*. Thanks to the phenomenological tradition, we know that this German word can be translated into French as *la chair* and into English as "the flesh." But we can go farther with the word *life*. Playing on the etymology, one might even be able to connect "life" with the French *lieu*, which would turn the flesh into a place. As Merleau-Ponty knew, the flesh is a half-way place (*un milieu*), the place of the interweaving, the place of the chiasm. In the nature lectures, Merleau-Ponty gives us an imperative that still holds in our present. It tells us how *not* to conceive an *archē* or principle in the epoch of anti-Platonism, in the epoch in which "immanence is complete." The principle must be conceived neither as positive nor as negative, neither as infinite nor as finite, neither as internal nor as external, neither as objective nor as subjective; it can be thought neither through idealism nor through realism, neither through finalism (or teleology) nor through mechanism, neither through determinism nor through indeterminism, neither through humanism nor through naturalism, neither through metaphysics nor through physics. Veering off into one of these extremes is precisely "what we must not do." In short, for Merleau-Ponty, there must be no *separation* between the two poles. But also, there must be no *coincidence*. Neither Platonism (separation) nor Aristotelianism (coincidence) is adequate. The positive formula for Merleau-Ponty's imperative would be the following: instead of either a separation or a coincidence, there must be "a hiatus," *un écart*, which *mixes* the two together. The mixture implies that, in fact, we are not really dealing with a principle, at least in its most traditional formulas: indivisible, sovereign, unified, a one. It is a principle without principle; the "principle" (the scare quotes are now necessary) is a paradox. It seems to me that, even today, fifty years later, the paradoxical mixture that Merleau-Ponty discovered, with its *minuscule hiatus*, still presents the question that calls thinking forth.

Merleau-Ponty's imperative, however, must be balanced with one from Deleuze: the ground must never *resemble* what it grounds. If the ground resembles, if it is copied off what it grounds, then we have used precisely what we are trying to explain in the explanation. Instead of resemblance, we must conceive the relation between the two poles, we must conceive the *écart*, as conflict. As conflict, the interweaving, "the folding over" (*le repli*) becomes "the unfold" (*le dépli*); the place becomes "spacing" (*espacement*); the milieu becomes "the nonplace" (*le non-lieu*). As conflict, the mixture becomes a battlefield: "the place of the dead" (*la place du mort*). As the place of the dead, the flesh is always virtually corrupt. As soon as we are born, we are ready to die. That's life. There is a powerlessness—originary finitude or finitization—within the "I can" of the flesh. This powerlessness puts blindness right in the middle of vision; it puts forgetfulness right in the middle of memory. The eye of representation, the "eye" of recognition, is gouged out. The powerless of vision, the powerless of memory, requires prosthetics, such as spectacles and writing. At the origin of life, we are always able to find technology; nature is always contaminated with culture (language again). Life is always out of joint. Since life is always out of joint, we have no sense or direction for making a jointure. All we can do is track the out(side) of the jointure (in a word, justice). By tracking it, by accompanying it, we become the ones who resist sense, who resist generalizations; we become the ones who resist valuation—as such *followers*, we become the ones who keep the promise. In order to resist valuation and keep the promise, we must hear the call to think; we must start to think. Thinking is a perilous act, since we cannot see clearly where we will end up. We might start to think that the singularity of a medical case—a life—demands that there be no law for its preservation. We might start to think that the multiplicity of a population demands that it be preserved in order to wander. We might start to think that the teeming presence of death that is in us, that is us—sui-cidal— demands that there be islands of life. Finally, we might start to think that the powerlessness of life demands that there be no pure filiations. It demands that there be neither light nor dark; powerlessness demands obscurity.

Notes

Introduction

1. "Specter" alludes to Jacques Derrida, *Spectres de Marx* (Paris: Galilée, 1993); English translation by Peggy Kamuf as *Specters of Marx* (New York: Routledge, 1994). In his eulogy for Gilles Deleuze, Derrida says, "When I was writing on Marx, at the very worst time, in 1992, I was somewhat reassured to find out that Deleuze intended to do the same thing." See Jacques Derrida, *The Work of Mourning*, ed. Pacale-Anne Brault and Michael Naas (Chicago: University of Chicago Press, 2001), p. 194.

2. See Michael Foucault, *Histoire de la sexualité, I: La Volonté de savoir* (Paris: Gallimard, 1976), esp. pp. 175–211; English translation by Robert Hurley as *The History of Sexuality, Volume I: An Introduction* (New York: Vintage, 1990), pp. 135–59.

3. For an interesting discussion of this case, see John Protevi, "The Schiavo Case: Jurisprudence, Biopower, and Privacy as Singularity," unpublished manuscript.

4. See Christian Caryl, "Why They Do It," in *The New York Review of Books* 52, no. 14 (September 22, 2005): 28–32.

5. See Avishai Margalit, "The Suicide Bombers," *The New York Review of Books* 50, no. 1 (January 16, 2003); online at www.nybooks.com/articles .15979. See also Dennis Keenan, *The Question of Sacrifice* (Albany: State University of New York Press, 2005).

6. My thanks to Sarah Clark Miller, who has provided essential insights into this kind of war rape. See Sarah Clark Miller, "Violence, Vulnerability and Agency: Understanding Genocidal Rape in Darfur," unpublished manuscript. See also Tara Gingerich and Jennifer Learning, "The Use of

Rape as a Weapon of War in the Conflict in Darfur, Sudan," Harvard School of Public Health: http://www.hsph.harvard.edu/fxbcenter/HSPH-PHR_Report_on_Rape_in _Dar fur.pdf. The Harvard report stresses the "ethnic cleansing" aim of war rape in the Darfur conflicts; see p. 18 in particular. This report also uses the terms *Arab* and *non-Arab* to distinguish between the groups. See also "The Crushing Burden of Rape: Sexual Violence in Darfur," a briefing paper by Médicins sans Frontières, Amsterdam, March 8, 2005: http://www.doctorswithoutborders.org/publications/re ports/2005/sudan03.pdf.

7. Gilles Deleuze, "Immanence: Une vie," in *Deux Régmies de fous* (1995; Paris: Minuit, 2003), pp. 359–63; English translation by Anne Boyman as "Immanence: A Life," in *Pure Immanence: Essays on a Life* (New York: Zone Books, 2001), pp. 25–33.

8. Michel Foucault, "Vie: Expérience et science," in *Dits et écrits, IV* (1984; Paris: Gallimard, 1994), pp. 763–76; English translation by Robert Hurley as "Life: Experience and Science," in *Essential Works of Michel Foucault: Aesthetics, Method, and Epistemology*, vol. 2, ed. James D. Faubion (New York: The New Press, 1998), pp. 465–78.

9. Jacques Derrida, "L'Animal que donc je suis (à suivre)," in *L'Animal autobiographique: Autour de Jacques Derrida* (Paris: Galilée, 1999), p. 285; English translation by David Wills as "The Animal That Therefore I Am (More to Follow)," in *Critical Inquiry* 28 (Winter 2002): 402. Derrida's lectures on animality were originally presented in 1998.

10. Derrida, "L'Animal," p. 300; "The Animal," p. 417.

11. Deleuze, "Immanence," p. 363; "Immanence," p. 31.

12. Foucault, "Vie," p. 776; "Life," p. 477. See also Thierry Hoquet's excellent introductory essay to *La Vie* (Paris: Flammarion, 1999), pp. 11–41, and *Notions de philosophie, I*, "Le Vivant" (Paris: Gallimard, 1995), pp. 231–300.

13. Martin Heidegger, *Gesamtausgabe*, vol. 9, *Wegmarken* (Frankfurt am Main: Klostermann, 1976), pp. 369–70 (pp. 199–200 of original publication); English translation edited by William McNeill as *Pathmarks* (Cambridge: Cambridge University Press, 1998), p. 281.

14. Gilles Deleuze, *Différence et répétition* (Paris: Presses Universitaires de France, 1968), p. 99; English translation by Paul Patton as *Difference and Repetition* (New York: Columbia University Press, 1994), p. 66.

15. Jacques Derrida, *De l'esprit: Heidegger et la question* (Paris: Galilée, 1987), pp. 175–76; English translation by Geoffrey Bennington and Rachel Bowlby as *Of Spirit: Heidegger and the Question* (Chicago: University of Chicago Press, 1989), pp. 106–7. Here we could also mention Levinas. Levinas claims that Heidegger is not able to think singularity because Heidegger starts from understanding and not the invocation of the other. But, as we shall see, Levinas plays no role in this book. We are still uncertain about Levinas's status. Can we say that in Levinas immanence is complete? See

Emmanuel Levinas, "Is Ontology Fundamental?" in *Basic Philosophical Writings*, ed. Adriaan T. Peperzak, Simon Critchley, and Robert Bernasconi (Bloomington: Indiana University Press, 1996), pp. 1–10, esp. 6.

16. The idea of a counter-science or counter-knowledge (*les savoirs assujettis*) comes from Foucault. See Michel Foucault, *"Il faut défender la société"*: *Cours au Collège de France, 1976* (Paris: Gallimard and Seuil, 1997), pp. 8–10; English translation by David Macey as *"Society Must Be Defended"*: *Lectures at the Collège de France, 1975–76* (New York: Picador, 2003), pp. 6–8.

17. Maurice Merleau-Ponty, *La Nature, notes cours du Collège de France* (Paris: Seuil, 1995), p. 203; English translation by Robert Vallier as *Nature: Course Notes from the Collège de France* (Evanston, Ill.: Northwestern University Press, 2003), p. 152.

18. Merleau-Ponty, *La Nature*, p. 164; *Nature*, p. 121.

19. This phrase comes from Michel Foucault, *Les Mots et les choses* (Paris: Gallimard, 1966), p. 351; anonymous English translation as *The Order of Things* (New York: Random House, 1970), p. 340. See also Jacques Derrida, *De la grammatologie* (Paris: Minuit, 1967), pp. 333–34; English translation by Gayatri Chakravorty Spivak as *Of Grammatology* (Baltimore: The Johns Hopkins University Press, 1976), p. 234. Here Derrida speaks of "une différence infime." One should also note that as recently as 2002 Derrida spoke of a "hiatus" (it is the same word in French) between "two equally rational postulations of reason." See Jacques Derrida, *Voyous* (Paris: Galilée, 2003), pp. 210–11; English translation by Pascale-Anne Brault and Michael Naas as *Rogues* (Stanford: Stanford University Press, 2004), p. 153.

20. Literally, this phrase means "the place of the dead." But it refers as well to the "dummy hand" in bridge. It is "the empty square" (*la case vide*). In Deleuze's essay on structuralism, the empty square is defined in the following way: "Games need the empty square, without which nothing would move forward or function. The object $= x$ is not distinguishable from its place, but it is characteristic of this place that it constantly displaces itself." Gilles Deleuze, "A quoi reconnaît-on le structuralisme?" in *L'Île déserte et autres textes* (Paris: Minuit, 2002), p. 261; English translation by Melissa McMahon as "How Do We Recognize Structuralism?" in *Desert Islands and Other Texts* (New York: Semiotext(e), 2004), p. 186. See also Jacques Derrida, "La Pharmacie de Platon," in *Dissemination* (Paris: Seuil, 1972), p. 104; English translation by Barbara Johnson as "Plato's Pharmacy" in *Dissemination* (Chicago: University of Chicago Press, 1981), p. 92.

1. *Verstellung* ("Misplacement")

1. Jean Hyppolite, *Logique et existence* (Paris: Presses Universitaires de France, 1952), p. 69; English translation by Leonard Lawlor and Amit Sen as *Logic and Existence* (Albany: State University of New York Press, 1997), p. 57. In following citations from this work, page numbers to the French edition will be given first, followed by the English.

2. Ibid., pp. 230/176.

3. Leonard Lawlor, *Derrida and Husserl: The Basic Problem of Phenomenology* (Bloomington: Indiana University Press, 2002). Hereafter abbreviated as DH.

4. See Leonard Lawlor, *Imagination and Chance: The Difference between the Thought of Ricœur and Derrida* (Albany: State University of New York Press, 1992).

5. Jacques Derrida, *Le Problème de la genèse dans la philosophie de Husserl* (Paris: Presses Universitaires de France, 1990), p. 226; English translation by Marian Hobson as *The Problem of Genesis in Husserl's Philosophy* (Chicago: University of Chicago Press, 2003). The translation here is my own.

6. Professor James Mensch has suggested that we might find thinking like that of Derrida elsewhere in the history of philosophy.

7. This paragraph is a partial response to Joshua Kates's review of *Derrida and Husserl* in *Husserl Studies* 21, no. 1 (April 2005): 55–64. In this paragraph I am trying to show how "this notion of a twofold critique, of an intra- and extra-phenomenological experience, which [I] presume applicable to all of Derrida's texts, related to genesis as the basic problem of phenomenology" (p. 61). See also Joshua Kates, *Essential History: Jacques Derrida and the Development of Deconstruction* (Evanston, Ill.: Northwestern University Press, 2005). Many of the points that Kates makes in the *Husserl Studies* review he makes again and in more detail in *Essential History*. After having noted, for instance, that I am "fairly sensitive to the specific historical variations that Derrida's thought overall may seem to undergo" (p. 229) and after noting that "it is to [my] credit to have raised the development issue explicitly at all" (p. 233), Kates claims that "Lawlor denegates many of the historical insights and actual differences in Derrida's thought that he himself often brings forward" (p. 235). In response to this claim, I must point out, again and nevertheless, that one of the primary projects of *Derrida and Husserl* consists in charting the shift in Derrida's thinking from a philosophy of the question to a philosophy of the promise. Kates also claims in *Essential History* that because I "see Derrida everywhere as a critic of phenomenology" (p. 235), I "short-circuit the question of the positive contribution phenomenology makes to Derrida's own thought" (p. 236). Here, I think that it is necessary to make three points. First, as I show in this paragraph, but also in the opening pages of *Derrida and Husserl* (esp. p. 3), phenomenology makes a positive contribution to Derrida's own thought, a positive contribution, in other words, to deconstruction: deconstruction always engages in a vigilance against metaphysical speculation that is not based on intuitive evidence. Second, in chapter 6 of *Derrida and Husserl*, which concerns Derrida's 1964 essay on Levinas, "Violence and Metaphysics," I say that "In 'Of Transcendental Violence,' besides criticizing Levinas for his unacknowledged presupposition of phenomenal presence, Derrida also *defends* [my emphasis; notice that I say "defends," not "criticizes"] Husserl from the

charges Levinas levels against his Fifth Cartesian Meditation" (p. 162); starting on the same page, I then lay out the structure of "Derrida's defense of Husserl." So it is plainly false that I "see Derrida *everywhere* as a critic of phenomenology" (my emphasis). Third, it seems necessary to point out that I have argued repeatedly in *Derrida and Husserl* that Derrida's thought *essentially depends* — I hope this "essentially depends" is clear enough — on Husserl's description of *Fremderfahrung* as *Vergegenwärtigung* in the Fifth Cartesian Meditation (see pp. 4, 135–38, 162–63, 174, 183–87, 194–95, 218, 231). When Kates discusses *Vergegenwärtigung* (p. 126 of *Essential History*), he does not cite *Derrida and Husserl*. But throughout *Derrida and Husserl* I argue that Derrida's thought is *precisely* a generalization of *Vergegenwärtigung* to all forms of experience, including and especially auto-affection. Although there are many ways to render this German term in English (or in French) due to the prefix *ver* — "representation," "depresentation," "presentification," "appresentation" — no matter what, *Vergegenwärtigung* implies a lack of full presence. In addition, I must point out that in my *Thinking through French Philosophy: The Being of the Question* (Bloomington: Indiana University Press, 2003) — Kates does cite this text — I say that there is no philosophy without the phenomenological reduction (pp. 148–51, esp. p. 149). For a different review of *Derrida and Husserl*, see Kas Saghafi, "Of Origins and Ends," in *Research in Phenomenology* 34 (2004): 303–14.

8. Jacques Derrida, *Positions* (Paris: Minuit, 1972), pp. 55–56; English translation by Alan Bass as *Positions* (Chicago: University of Chicago Press, 1981), pp. 41–42.

9. Martin Heidegger, "Die Sprache," in *Unterwegs zur Sprache* (Pfullungen: Neske, 1959), p. 12; English translation as "Language" by Albert Hofstadter in *Poetry, Language, Thought* (New York: Harper Collophon, 1971), p. 190: "Die Sprache selbst ist die Sprache." Derrida explicitly refers to this monologue in *The Monolingualism of the Other*. See Jacques Derrida, *Le Monolinguisme de l'autre: ou la prothèse d'origine* (Paris: Galilée, 1996), p. 129; English translation by Patrick Mensah as *Monolingualism of the Other; or, the Prosthesis of Origin* (Stanford: Stanford University Press, 1998), p. 69.

10. Martin Heidegger, *Wegmarken* (Frankfurt am Main.: Klostermann, 1967), p. 369; English translation by William McNeill as *Pathmarks* (Cambridge: Cambridge University Press, 1998), p. 284. I have modified the English translation.

11. Michel Foucault, "La Pensée du dehors," in *Dits et écrits, I* (Paris: Gallimard Quarto, 2001), pp. 546–67; English translation by Brian Massumi as "The Thought from Outside," in *Foucault / Blanchot* (New York: Zone Books, 1997), pp. 7–60. For an excellent discussion of this text, see Kas Saghafi, "The 'Passion for the Outside': Foucault, Blanchot, and Exteriority," in *International Studies in Philosophy* 28, no. 4 (1996): 80–92.

12. Leonard Lawlor, *Thinking through French Philosophy: The Being of the Question* (Bloomington: Indiana University Press, 2003). I am about to

begin a new book, which will be a kind of "prequel" to *Thinking through French Philosophy*. It is to be called *Continental Philosophy before 1960: Toward the Outside* (Indiana University Press, forthcoming). It will present a philosophical history, through Bergson, Freud, Husserl, Heidegger, and Merleau-Ponty, that reaches its culmination in the thought from the outside.

13. Jacques Derrida, *La Voix et le phénomène* (Paris: Presses Universitaires de France, 1967), p. 20; English translation by David B. Allison as *Speech and Phenomena* (Evanston, Ill.: Northwestern University Press, 1973), p. 21. Hereafter cited as VP with reference first to the French, then to the English translation. I will translate the title *Voice and Phenomenon*, however; for justification, see DH.

14. Michel Foucault, *Les Mots et les choses* (Paris: Gallimard, 1966), p. 336; anonymous English translation as *The Order of Things* (New York: Random House, 1970), p. 325. Hereafter cited as MC, with reference first to the French, then to the English translation, which I have modified in the quotes I have reproduced here. In the text, I have translated the French title more literally, in order to underline parallels with other thinkers, Derrida in *Voice and Phenomenon* especially.

15. See Chapter 10 below, esp. the first two sections.

16. Derrida makes a similar comment, stressing form and content, in the Introduction to *Voice and Phenomenon*: presence has always been and will always be, to infinity, the form in which—we can say this apodictically—the infinite diversity of content will be produced. The opposition—which inaugurates metaphysics—between form and matter finds in the concrete ideality of the living present its ultimate and radical justification (VP 5/6).

17. Gilles Deleuze, *Foucault* (Paris: Minuit, 1986), p. 137n10; English translation by Seán Hand as *Foucault* (Minneapolis: University of Minnesota Press, 1988), p. 152n10. Cf. also MC 291/278.

18. Jacques Derrida, "The Animal That Therefore I Am (More to Follow)," translated by David Wills, in *Critical Inquiry* 28 (Winter 2002): p. 402. My thanks to Brett Buchanan for alerting me to this text. See also p. 392, where Derrida speaks of *un autre vivant*. This and the other extant portions of the book Derrida intended to write on the human/animal distinction, left unfinished at his death, have been published in French as *L'Animal que done je suis*, ed. Marie-Louise Mallet (Paris: Galilée, 2006); English translation by David Wills forthcoming from Fordham University Press.

19. See chapter 3 of this volume.

20. Edmund Husserl, Hua III.1: *Ideen zu einer reinen Phänomenologie und phänomenologischen Philosophie*, bk. 1, ed. Karl Schuhmann (The Hague: Martinus Nijhoff, 1976); English translation by Fred Kersten as *Ideas pertaining to a Pure Phenomenology and to a Phenomenological Philosophy* (The Hague: Martinus Nijhoff, 1982). See also Edmund Husserl, *Idées directrices pour une phénoménologie*, trans. Paul Ricœur (Paris: Gallimard, 1950).

21. Here I am relying on Husserl's later revision of the passage: "copy D." See Kersten's English translation, p. 73.

22. *Reelle* is a technical term that means transcendental (not psychological); it is a part of psychic life, but psychic life not considered as a part of nature, as it is studied by the science of psychology.

23. This solution to the transcendental problem, a solution that defines "psychologism," is circular because it takes something existing in the world, the psyche, which has the ontological sense of something existing in the world, *Vorhandenheit*, and tries to make this something present account for all things present.

24. Edmund Husserl, Hua IX: *Phänomenologische Psychologie* (The Hague: Martinus Nijhoff, 1962), p. 292; English translation by Richard E. Palmer in *The Essential Husserl*, ed. Donn Welton (Bloomington: Indiana University Press, 1999), p. 331.

25. Ibid., pp. 294 / 332.

26. Here is indeed the divergence, *l'écart*, between phenomenology and nonphenomenology (deconstruction). See Steven Galt Crowell, *Husserl, Heidegger, and the Space of Meaning* (Evanston, Ill.: Northwestern University Press, 2001), p. 172.

27. Michel Foucault, *L'Ordre du discours* (Paris: Gallimard, 1971), p. 80; English translation by Alan Sheridan as "The Discourse on Language," in *The Archeology of Knowledge* (New York: Pantheon, 1972), p. 237.

28. Ibid., pp. 72 / 234.

2. With My Hand over My Heart, Looking You Right in the Eyes, I Promise Myself to You . . .

1. Frédéric Worms is preparing a book on the philosophical moments of the twentieth century, one of which will be "le moment 1968."

2. Edmund Husserl, *Ideen zu einer reinen Phänomologie und phänomenologischen Philosophie*, bk. 1, *Allgemeine Einführung in die reine Phänomenologie*, ed. Karl Schuhmann (The Hague: Martinus Nijhoff, 1976) (Hua III.1, 2); English translation by Fred Kersten as *Ideas pertaining to a Pure Phenomenology and to a Phenomenological Philosophy. First Book: General Introduction to a Pure Phenomenology* (The Hague: Martinus Nijhoff, 1983).

3. Jacques Derrida, *Le Toucher—Jean-Luc Nancy* (Paris: Galilée, 2000). Hereafter cited as LT; all English translations are my own.

4. In this essay, focusing primarily on "Tangent II," I have underrepresented the immense role that Nancy's thought plays in this book. Indeed, the entire book is a reflection on Nancy's corpus. Perhaps we should say that *Le Toucher* is a work of gratitude since, thanks to Nancy's heart transplant, Derrida did not have to see yet another one of his friends die before he did (cf. LT 13).

5. For more on Derrida's eschatology in relation to Foucault's critique of phenomenology, see chap. 4 of this volume.

6. Edmund Husserl, *L'Origine de la géométrie*, trans. and introd. Jacques Derrida (1962; Paris: Presses Universitaires de France, 1974); English

translation by John P. Leavey, Jr., as *Edmund Husserl's Origin of Geometry: An Introduction* (1978; Lincoln: University of Nebraska Press, 1989). Hereafter cited as LOG.

7. Jacques Derrida, "'Genèse et structure' et phénoménologie," in *L'É-criture et la différence* (Paris: Seuil, 1967); English translation by Alan Bass as *Writing and Difference* (Chicago: University of Chicago Press, 1978). Hereafter cited as ED. For the peculiar genesis of this essay, see my *Derrida and Husserl: The Basic Problem of Phenomenology* (Bloomington: Indiana University Press, 2000), esp. p. 29.

8. This comment anticipates what we will see in *Le Toucher*, where Derrida will employ the verb *faire partie* in his argumentation.

9. This non–self-identity is a "trace." The trace implies that there is a repetition prior to any present impression. For Derrida, we can say in the most general terms that memory precedes perception. A memory in turn implies that there is a lateness and therefore distance from what is remembered, which is no longer present. The primacy of memory already anticipates Derrida's discourse of the heart: "learning by heart."

10. The second part of the book is called "Exemplary Histories of the 'Flesh.'"

11. Edmund Husserl, *Ideen zu einer reinen Phänomenologie und phänomenologischen Philosophie*, bk. 2, *Phänomenologische Untersuchungen zur Konstitution*, ed. Marly Biemal (The Hague: Martinus Nijhoff, 1952) (Hua IV); English translation by Richard Rojcewicz and André Schuwer as *Ideas pertaining to a Pure Phenomenology and to a Phenomenological Philosophy*, bk. 2, *Studies in the Phenomenology of Constitution* (Dordrecht: Kluwer Academic Publishers, 1989).

12. "Tangent II" can be divided into three parts: pp. 184–94, which concern §37 and the privilege of digital touching; pp. 194–201, which still concern §37, though now the focus is on the difference between the eye and the hand; and finally pp. 201–6, which concern §45 and the feeling of the heart. P. 183 is an introductory page on exemplarity, and pp. 206–8 form a conclusion to the chapter.

13. As we noted above, *Leib* is commonly translated into French as *la chair* ("the flesh"); Derrida renders it as "le corps propre" ("one's own body").

14. Here is the passage in its entirety (Hua IV.147; English translation, p. 155): "(1) In the tactile realm we have the external object, tactually constituted, and a second object, the body, likewise tactually constituted, for example, the touching finger, and, in addition, there are fingers touching fingers. (2) So here we have that double apprehension: the same touch-sensation is apprehended as a feature of the 'external' object and is apprehended as a sensation of the body as object. (3) And in the case in which a part of the body becomes equally an external object of another part, we have the double sensation (each part has its own sensations) and the double

apprehension as feature of the one or the other bodily part as a physical object." Derrida inserts the numeration, and "body" here renders *Leib*.

15. This is on p. 155 of the English translation.

16. This *es fehlt* (Hua IV.148) is rendered as "what is denied" on pp. 155–56 of the English translation.

17. For the importance of this idea of lack in Derrida, Deleuze, and Foucault, see my *Thinking through French Philosophy* (Bloomington: Indiana University Press, 2003).

18. In this quotation, Derrida is using the French idiom *à même*, which I have rendered as "right on." This idiom is complicated, since it involves both the preposition *à* and the adverb or noun *même*; taken together, the two words imply a mixture of immediacy (*même*: the same) and mediation (*à*: the indirection of the address to). With this idiom, Derrida is suggesting the foreignness of the "local coincidence" that is important for Husserl in the double sensation of touching and being touched. For more on this idiom, see my *Thinking through French Philosophy*.

19. This quotation appears on p. 149 of Hua IV and p. 157 of the English translation.

20. "There is good reason to chose the 'example' of the hand as the starting point in the analysis of the 'external object that is constituted in a tactile way.' It is that a certain exteriority, an exteriority that is heterogeneous to the sensible impression that is real (and, as Husserl recalls, that is an optical property of the hand), *participates*, a perceived exteriority *as* real *must* participate in the experience of touching and being touched, as well as in the double apprehension, be it by virtue of the hyle-morphe relation or by virtue of the noetic-noema relation, and even in the case of illusion. The duplicity of this apprehension would not be possible unless an outside, with its real, thinglike quality, announced itself in the sensible impression, and already in its hyletic content. This exteriority is necessary. This foreign outside is necessary, foreign at once to the touching side and to the touched side of the phenomenological impression, where the latter is not given in adumbrations" (LT 200).

21. In "Tangent IV" (on Didier Franck), Derrida also speaks of Husserl, of "what, *phenomenologically*, *suspends* the contact *in* contact and divides it right on tactile experience in general" (LT 257, Derrida's emphasis). More importantly, Derrida is saying this as if he were arguing with Nancy and Franck. In other words, he is presenting his own position by means of the phenomenological viewpoint. Here we can see the continuity with *Voice and Phenomenon*. This argument encapsulates Derrida's entire thought, at least in *Le Toucher*, if not in all of his work; it encapsulates his critique of phenomenology. Again, adopting the phenomenological viewpoint, Derrida invokes "the double motif that is classic in Husserl" (LT 257). On the one hand, there is the epochē. Its very possibility consists in suspending the "reality" of contact in order to deliver its intentional or phenomenal sense. Derrida

says, "the *sense* of the contact is given to me, as such, through this interruption or through this suspensive conversion. I cannot therefore have, fully, *both* contact *and* the sense of contact" (LT 257, Derrida's emphasis). The important word in this comment is *fully* (*pleinement*), which implies that there is no full intuition of either contact or the sense of contact. But, Derrida continues: "the noematic content . . . can appear, it can phenomenalize itself only by *not* belonging *really* [réellement] *either* to the thing touched . . . *or* to the material of my *Erlebnis*" (LT 257; Derrida's emphasis). There is no full intuition because the noema is *irreell*, neither *reell* in the sense of belonging to my lived-experience nor *real* in the sense of belonging to the thing touched. This, as Derrida says, "double possibility (epochē and non-*reell*-belonging of the content of the intentional sense)" (LT 257) would open up a nonadherence of the noema of what is touched to my lived-experience or to the thing. Not belonging to either region of the tactile experience, the noema inhabits a different space, "the spacing of a distance." And this spacing implies mediation, understood as interruption, nonsimultaneity, *Vergegenwärtigung*. In this context, Derrida calls this spacing of the noema *différance*. But there is more. Because the spacing introduces some sort of mediation, Derrida can claim that there is both nonlife in *Erlebnis* and a kind of "technological prosthetic": "'technology'," he says, "is thus called for by the phenomenological necessity itself" (LT 258).

22. This passage occurs in Hua IV.149 and on p. 157 of the English translation.

23. Derrida is commenting on the long description that runs across Hua IV.165–66, which is pp. 173–74 of the English translation. Derrida in fact divides the description into three phases.

24. The French translation renders Husserl's *durch* as *à travers*; the English translation renders it as *through*.

25. I have coined a word for this idea of "a promise that demands to be done over again and again": *refinition*. See the preface to my *Derrida and Husserl*, also *Thinking through French Philosophy*.

26. In French, the question is: "Quand nos yeux se touchent, fait-il jour ou fait-il nuit?" Derrida claims to have seen this question written on a wall in Paris (LT 13–14).

3. "For the Creation Waits with Eager Longing for the Revelation"

1. While the phrase "otherwise than being" alludes to Levinas, we find a similar expression in Derrida: "a living without being." See Jacques Derrida, *Sovereignties in Question: The Poetics of Paul Celan*, ed. Thomas Dutoit and Outi Pasanen (New York: Fordham University Press, 2005), pp. 110 and 126.

2. Jacques Derrida, "L'Animal que donc je suis (à suivre)," in *L'Animal autobiographique: Autour de Jacques Derrida* (Paris: Galilée, 1999), p. 285; English translation by David Wills as "The Animal That Therefore I Am

(More to Follow)," in *Critical Inquiry* 28 (Winter 2002): 402. It is important that Derrida, in this quote from "The Animal That Therefore I Am," speaks of "the living" (*le vivant*) and not "lived-experience" (*le vécu*). *Le vécu* is the French translation of the German term *Erlebnis*, which was central to all classical phenomenological investigations. Without reconstructing the complex descriptions of lived-experience that we can find in Husserl, we need think only of Descartes's *ego cogito* in order to get a sense of what this concept is about. Lived-experience refers to the experience of being conscious of one's self; it connotes, therefore, self-presence and self-coincidence, simultaneity and lucidity; it is an auto-affection, which is tautological, *to auto*, the same. The concept of *Erlebnis* therefore is based on the simplest and most traditional definition of life—life as auto-affection, as sensibility or irritability.

3. See, e.g., Hans Jonas, *The Phenomenon of Life* (New York: Harper and Row, 1966), p. 99. See also Derrida, "The Animal That Therefore I Am (More to Follow)," 300 / 417.

4. In *De Anima*, Aristotle also says that the soul is, in a sense, all existing things (431b21). Heidegger, of course, cites this passage in *Being and Time*. See Martin Heidegger, *Sein und Zeit* (Tübingen: Niemeyer, 1979), p. 14; English translation by Joan Stambaugh as *Being and Time* (Albany: State University of New York Press, 1996), p. 17.

5. One should recall that Foucault, too, mentions this fact. See MC 22 / 6.

6. For the use of this term, see LT, esp. pp. 258 and 318.

7. See, e.g., Jacques Derrida, *Le Monolinguisme de l'autre* (Paris: Galilée, 1996), pp. 122–23; English translation by Patrick Mensah as *Monolingualism of the Other; or, the Prosthesis of Origin* (Stanford: Stanford University Press, 1998), pp. 64–65.

8. Jacques Derrida, *Mémoires d'aveugle: L'Autoportrait et autre ruines* (Paris: Editions de la Réunion des musées nationaux, 1990); English translation by Pascale-Anne Brault and Michael Naas as *Memoirs of the Blind: The Self-Portrait and Other Ruins* (Chicago: University of Chicago Press, 1993). Hereafter cited as MdA with reference first to the French, then to the English. *Memoirs of the Blind* is an occasional text, like so many of Derrida's works. In fact, it is presented as a sort of interview. Apparently, there was a video made in which a woman interrogated Derrida about the drawings. See Michael Newman, "Derrida and the Scene of Writing," *Research in Phenomenology* 24 (1994): 218–34.

9. The following texts have been consulted in writing this essay: Michael Newman, "Derrida and the Scene of Writing," *Research in Phenomenology* 24 (1994): 218–34; Michael Kelly, "Shadows of Derrida," *Artforum* 29, no. 6 (1991): 102–4; Meyer Raphael Rubinstein, "Sight Unseen," *Art in America* 79, no. 4 (April 1991): 47–51; Michael Fried, *Manet's Modernism* (Chicago: University of Chicago Press, 1996), esp. pp. 365–73. Both Kelly

and Rubenstein note the irony that the first picture seen in this exhibition of drawings was not a drawing but a painting, *Butades or the Origin of Drawing*. But, as I try to show here, the reason for this placement is obvious, since it gives us a self-portrait of painting or drawing. Neither Rubenstein nor Kelly nor Fried mentions that Jan Provost's painting *Sacred Allegory* was number 36 of the 44 pictures in the exhibition; in fact, they do not mention this painting at all. Fried is interested in *Memoirs of the Blind* only because of Derrida's discussion of Fantin-Latour. I have also consulted: John D. Caputo, *The Prayers and Tears of Jacques Derrida* (Bloomington: Indiana University Press, 1997); Robert Vallier, "Blindness and Invisibility: The Ruins of Self-Portraiture (Derrida's Re-Reading of Merleau-Ponty)," in *Écart and Différance*, ed. Martin C. Dillon (New Jersey: Humanities Press, 1997), pp. 191–207; and Véronique Fóti, "The Gravity and the (In)visibility of the Flesh: Merleau-Ponty, Nancy, Derrida," in her *Vision's Invisibles* (Albany: State University of New York Press, 2003), pp. 69–80. I first read *Memoirs of the Blind* carefully in 1998 at the Collegium Phaenomenologicum, Columbello, Umbria, where Michael Newman was lecturing on it.

10. See Jacques Derrida, "Violence and Metaphysics," in ED 227 / 153. I had already characterized deconstruction in Derrida as wider in DH (p. 148). For more on the "jewgreek," see Jacques Derrida, *Force de la loi* (Paris: Galilée, 1994), pp. 131–32; English translation by Mary Quaintance as "Force of Law," in *Deconstruction and the Possibility of Justice*, ed. Drucilla Cornell, Michael Rosenfeld, and David Gary Carlson (New York: Routledge, 1992), p. 56.

11. On the question of "wider," it is important to recall that Heidegger, in *Being and Time*, says that being is "the *transcendens* pure and simple" (*Being and Time*, trans. Joan Stambaugh [Albany: State University of New York Press, 1996], 33). He comments on this in the "Letter on Humanism": "Just as the openness of spatial nearness seen from the perspective of a particular thing exceeds all things near and far, so is Being essentially wider [*weite*] than all beings, because it is the clearing itself." See Martin Heidegger, "Brief über den Humanismus," in *Wegmarken* (Frankfurt am Main: Klostermann, 1978), p. 333; English translation by Frank A. Capuzzi in collaboration with J. Glenn Gray as "Letter on Humanism," in *Martin Heidegger: Basic Writings*, ed. David Farrell Krell (New York: Harper Collins, 1993), p. 240.

12. In reference to this claim, it is important to keep in mind that a deconstruction of Christianity would disclose an origin that is non-Christian. The non-Christian origin therefore could be called Jewish, Greek, or Islamic. A deconstruction of Christianity would precisely attempt to move back from Christianity to its sources in these other religions. In fact, without a deconstruction, Judaism and Islam could be ignored as important components of the West. The deconstruction of Christianity would focus, it seems to me, on its catholic nature, which would raise issues of boundaries. It

would also focus on the event of the Christ, God becoming incarnated, made flesh.

13. Caputo, *The Prayers and Tears of Jacques Derrida*, has been particularly helpful in the writing of this essay, especially the chapters on *Circumfession* and *Memoirs of the Blind* (pp. 281–329). Caputo's is the only text that focuses, briefly, on the Provost painting. Despite the help that this book provides, Caputo's interpretation of *Memoirs of the Blind* seems to be too one-sided, insofar as at one point he describes Derrida's thought as a "philosophy of blindness" (p. 318). This claim does not fit well with Derrida's deconstruction of phenomenology in *Voice and Phenomenon*, which uses intuitionism against formalism and vice versa. Also, in *Memoirs of the Blind* Derrida repeatedly speaks of the *usure* of the eyes, which implies a using up of the eyes to the point of blindness but also the usury of the eyes in order to see too much. Finally, one must recognize the association between tears and writing, and writing, of course, is to be read, with the eyes.

14. The following section is a reading of MdA 9–61 / 1–57, esp. 46–61 / 41–57.

15. The painting, on loan to the Louvre for this exhibition, depicts a woman, Butades, drawing her lover by outlining his shadow on a wall. It appears as fig. 19 in MdA.

16. See also Jacques Derrida, *De la grammatologie* (Paris: Minuit, 1967), pp. 333–34; English translation by Gayatri Chakravorty Spivak as *Of Grammatology* (Baltimore: The Johns Hopkins University Press, 1976), p. 234. Discussing Rousseau, Derrida describes the origin of drawing as it is depicted in this painting.

17. See MC 351 / 340, and chap. 5 of the present book. For a recent use of the word *écart* in Derrida, see LT 322.

18. For more on *faute*, see Jacques Derrida, "En ce moment même . . . ," in *Psyché: Inventions de l'autre* (Paris: Galilée, 1987), p. 164; English translation by Ruben Berezdivin as "At This Very Moment in This Work Here I Am," in *Re-Reading Levinas*, ed. Robert Bernasconi and Simon Critchley (Bloomington: Indiana University Press, 1991), p. 15.

19. Derrida, *Of Grammatology*, 297–98 / 208–9.

20. In "Parergon" (first published in 1974), Derrida calls this separation "the paradox of [Kant's] Third Critique": dealing with singularities that must give rise to universal judgment. See Jacques Derrida, *La Vérité en peinture* (Paris: Flammarion, 1978), p.106; English translation by Geoff Bennington and Ian McLeod as *The Truth in Painting* (Chicago: University of Chicago Press, 1987), p. 93.

21. This absence implies, as we will see, in the case of religious paintings, that God is dead. See also Gilles Deleuze, *Logique de la sensation* (Paris: Edition de la difference, 1981), pp. 13–14, where Deleuze speaks of a "pictorial atheism."

22. The form would be what Derrida calls in *The Truth in Painting* a "paradigm" (cf. MdA 64 / 60); literally, a paradigm is "beyond showing" (*para-*

deiknynai). See Derrida, *The Truth in Painting*, pp. 223–24 / 194–95. Apparently, given the roots of the word *paradigm*, Derrida intends with this term a model that is beyond (*para*) showing (*deiknynai*). Derrida also connects this term to the discourse of negative theology in "Comment ne pas parler, Dénégations," in *Psyché: Inventions de l'autre*, p. 582; English translation by Ken Frieden as "How to Avoid Speaking: Denials," in *Languages of the Unsayable*, ed. Sanford Budick and Wolfgang Iser (New York: Columbia University Press, 1989), p. 50.

23. The following section concerns MdA 61–96 / 57–92.

24. In *Memoirs of the Blind*, Derrida repeatedly speaks of hypotheses, but especially the hypothesis of sight: see MdA 1 / 1, 9–10 / 2–3, 58 / 54, 64 / 60, 119 / 117. I have not discussed the hypothesis of sight, since every text that has been written about *Memoirs of the Blind* mentions it. Probably Michael Fried summarizes it best (*Manet's Modernism*, p. 367): "Derrida's point concerns the inescapableness of hypothesizing (of conjecturing, presupposing) at the heart of the act of intuiting (that is, of seeing, as it were immediately and without reflection)."

25. Derrida's considerations of the self-portrait, into which we are about to enter, refer back, of course, to Lacan's discussion of the mirror stage. See Jacques Lacan, "Le Stade du mirroir," in *Écrits* (Paris: Seuil, 1966), pp. 93–100; English translation by Alan Sheridan as "The Mirror Stage," in *Ecrits* (New York: Norton, 1977), pp. 1–7. See also Jacques Derrida, "And Say the Animal Responded," in *Zoontologies*, ed. Cary Wolfe (Minneapolis: University of Minnesota Press, 2003). It also alludes to Merleau-Ponty, who appears in *Memoirs of the Blind* (MdA 56–58 / 52–53), especially to the discussions of the mirror that we find in "Eye and Mind." See Maurice Merleau-Ponty, *L'Œil et l'esprit* (Paris: Gallimard, 1964), esp. 19–20 and 32–33. For an excellent reading of Derrida's *Memoirs of the Blind* in relation to Merleau-Ponty, see Vallier, "Blindness and Invisibility." In a note, Vallier recommends consulting Michael Fried's *Courbet's Realism* (Chicago: University of Chicago Press, 1990) in order to learn more about the self-portrait. Indeed, chap. 2 of *Courbet's Realism* is particularly interesting, as Vallier predicts; see esp. pp. 66–67, where Fried speculates that sleeping (a nonvertical position) may be the primordial relation to the world, and pp. 78–84, where Fried claims that the self-portrait in Courbet concerns self-identification. And, finally, Derrida's considerations resemble—we cannot ignore this—a description of a certain painting by Velázquez called in French *Les ménines*. See MC chap. 1. See also Gary Shapiro, *Archeologies of Vision* (Chicago: University of Chicago Press, 2003).

26. See Fried, *Manet's Modernism*, p. 370.

27. See Jacques Derrida, *Spectres de Marx* (Paris: Galilée, 1993), pp. 28–29, 164; English translation by Peggy Kamuf as *Specters of Marx* (New York: Routledge, 1994), pp. 8–9, 100.

28. See Derrida, "Force of Law," "And then I would love to write . . . a short treatise on love of ruins" (p. 44).

29. See also Jacques Derrida, *Adieu à Emmanuel Levinas* (Paris: Galilée, 1997), p. 121; English translation by Pascale-Anne Brault and Michael Naas as *Adieu to Emmanuel Levinas* (Stanford: Stanford University Press, 1999), p. 67.

30. The following section interprets MdA 96–130 / 92–127. For a very interesting discussion of conversion in *Memoirs of the Blind*, indeed, a very interesting interpretation of *Memoirs of the Blind* as a whole, see Michael Naas and Pascale-Anne Brault, "Better Believing It," in Michael Naas, *Taking on the Tradition: Jacques Derrida and the Legacies of Deconstruction* (Stanford: Stanford University Press, 2003), pp. 117–35, esp. p. 126.

31. Heidegger also speaks of this desire of the eyes in Augustine in *The Phenomenology of Religious Life* (Bloomington: Indiana University Press, 2004).

32. The painting is reproduced as fig. 69 of MdA. One can also view this painting at the Louvre Web site: www.louvre.fr / louvrea.htm.

33. For more on allegory, see Derrida, "Psyché: Inventions de l'autre," in *Psyché*, pp. 20–21.

34. Concerning love in Derrida, see John Protevi, "Love," in *Between Deleuze and Derrida*, ed. Paul Patton and John Protevi (New York: Continuum, 2003), pp. 183–94. See also Jacques Derrida, *Politiques de l'amitié* (Paris: Galilée, 1994); English translation by George Collins as *Politics of Friendship* (New York: Verso, 1997).

35. See Derrida, "How to Avoid Speaking," pp. 572–74 / 41–42.

36. Jacques Derrida, "Circonfession," in Geoffrey Bennington and Jacques Derrida, *Jacques Derrida* (Paris: Seuil, 1991), p. 62; English translation by Geoffrey Bennington as "Circumfession," in *Jacques Derrida* (Chicago: University of Chicago Press, 1993), p. 62.

37. See ibid., 12 / 10.

38. Jacques Derrida, " 'Il faut bien manger' ou le calcul du sujet," in *Points de suspension* (Paris: Galilée, 1992), p. 288; English translation by Peter Connor and Avital Ronell as " 'Eating Well,' or the Calculation of the Subject," in *Points . . .: Interviews, 1974–1994* (Stanford: Stanford University Press, 1995), p. 274. See also Jacques Derrida, *Résistances de la psychanalyse* (Paris: Galilée, 1996); English translation by Peggy Kamuf, Pascale-Anne Brault, and Michael Naas as *Resistances of Psychoanalysis* (Stanford: Stanford University Press, 1998). But the clearest statement Derrida makes concerning analysis comes from his 1968 essay "Plato's Pharmacy": "All translations into languages that are the heirs and depositaries of Western metaphysics thus produce on the pharmakon an effect of analysis that violently destroys it, reduces it to one of its simple elements by interpreting it, paradoxically enough, in the light of the ulterior developments it itself has made possible." See "La Pharmacie de Platon," in *La Dissémination* (Paris: Seuil, 1972), p. 112; English translation by Barbara Johnson as "Plato's Pharmacy," in *Dissemination* (Chicago: University of Chicago Press, 1981), p. 99.

39. Throughout *Le Toucher*, Derrida refers or alludes to Jean-Luc Nancy's project of a "deconstruction of Christianity." See, e.g., LT 68. For Nancy's own statement of this project, see Jean-Luc Nancy, "La Déconstruction du christianisme," *Les Etudes Philosophiques*, no. 4 (1998): 503–19; English translation by Simon Sparks as "The Deconstruction of Christianity," in *Religion and Media*, ed. Hent de Vries and Samuel Weber (Stanford: Stanford University Press, 2001), 112–30. The essay is included in Nancy, *La Déclosion: Déconstruction du christianisme, 1* (Paris: Galilée, 2005), pp. 203–26. Another portion of this project is Jean-Luc Nancy, *Visitation (de la peinture chrétienne)* (Paris: Galilée, 2001); English translation by Jeff Fort as "Visitation: Of Christian Painting," in *The Ground of the Image* (New York: Fordham University Press, 2005), pp. 108–25.

40. It is perhaps not surprising that Nancy, in "The Deconstruction of Christianity," connects this project back to the thought of Derrida. Nancy says, "might one not wonder whether the 'JewGreek' of which Derrida speaks at the end of 'Violence and Metaphysics,' this 'JewGreek' that is our history, is not, in fact, Christian?" (p. 504 / 113).

41. LT 299–300.

42. Nancy, "The Deconstruction of Christianity," p. 507 / 116.

43. Nancy calls the belief that the origin of Christianity is unified "the 'projection of Christmas,'" ibid., p. 509 / 118.

44. This is the one place in *Memoirs of the Blind* where Derrida speaks of negative theology. For more on negative theology, see Thomas Carlson, *Indiscretion: Finitude and the Name of God* (Chicago: University of Chicago Press, 1999).

45. Derrida, *Adieu to Emmanuel Levinas*, pp. 126 and 204 / 69 and 119.

46. Jacques Derrida, *Shibboleth* (Paris: Galilée, 1986), p. 111; English translation by Joshua Wilner as "Shibboleth," in *Sovereignties in Question: The Poetics of Paul Celan*, ed. Thomas Dutoit and Outi Pasanen (New York: Fordham University Press, 2005), p. 62.

47. See Derrida, "Circumfession," pp. 146–47 / 155–56.

48. Derrida, *Adieu to Emmanuel Levinas*, pp. 181–82 / 104–5.

49. Nancy, "The Deconstruction of Christianity," p. 511 / 121.

50. Indeed, *Le Toucher* is largely an extended engagement with Jean-Luc Nancy, *Corpus* (Paris, A. M. Métailié, 1992); English translation by Richard Rand forthcoming from Fordham University Press.

51. See Martin Heidegger, *Nietzsche*, 2 vols. (Pfullingen: Neske, 1961), 2:199–202; English translation by David Farrell Krell as *Nietzsche: Volume Four, Nihilism* (New York: Harper and Row, 1982), pp. 147–49. See also Martin Heidegger, "Nietzsches Wort 'Gott ist tot,'" in *Holzwege* (Frankfurt am Main: Klostermann, 1957), pp. 193–247; English translation by William Lovitt as "Nietzsche's Word 'God is Dead,'" in *The Question concerning Technology and Other Essays* (New York: Harper and Row, 1977), pp. 53–113.

52. Heidegger, *Nietzsche*, 1:544; English translation by David Farrell Krell as *Nietzsche: Volume Three, The Will to Power as Knowledge and as Metaphysics* (New York: Harper and Row, 1987), p. 61.

53. Ibid., 1:623 / p. 128.

54. Heidegger, "Nietzsche's Word 'God is Dead,'" p. 210 / 71. See also Heidegger, *Nietzsche*, 2:102; *Volume Four*, p. 63.

55. Heidegger, "Nietzsche's Word 'God is Dead,'" pp. 242–43 / 108.

56. One should add that Deleuze, in *Nietzsche and Philosophy*, starts with a discussion of value in Nietzsche. See *Nietzsche et la philosophie* (Paris: Presses Universitaires de France, 1962), pp. 1–2; English translation by Hugh Tomlinson as *Nietzsche and Philosophy* (New York: Columbia University Press, 1983), pp. 1–2. Here Deleuze connects valuation in Nietzsche to what he calls "the differential element." In 1972, in "A quoi reconnaît-on le structuralisme?" (in *L'Île déserte et autres texts* [Paris: Minuit, 2002], pp. 238–69; English translation by Michael Taormina, ed. David Lapoujade, as *Desert Islands and Other Texts, 1953–1974* [New York: Semiotex(te), 2004], p. 186), Deleuze describes "the differential element" as a "blind spot" (see p. 261); we should also note that Deleuze says this in reference to Foucault's description of the Velázquez painting. An English translation of this essay can be found in Charles Stivale's *Post-texts*, p. 275.

57. See Derrida, "The Animal That Therefore I Am," p. 278 / 396.

58. Jacques Derrida, *Voyous* (Paris: Galilée, 2003), p. 205; English translation by Pacale-Anne Brault and Michael Naas as *Rogues* (Stanford: Stanford University Press, 2005), p. 149.

59. In regard to this blindness, one should recall that Heidegger says that the original meaning of the word *Ereignis* is *er-äugen*, to place before the eyes. See Martin Heidegger, *Identität und Differenz* (Pfullingen: Neske, 1957), pp. 24–25; English translation by Joan Stambaugh as *Identity and Difference* (New York: Harper and Row, 1969), pp. 100–101. This claim is left untranslated in the English version. See also Françoise Dastur, "Phénoménologie de l'événement," in *Phénoménologie en questions* (Paris: Vrin, 2004), pp. 172–73.

60. Jacques Derrida, *Sauf le nom* (Paris: Galilée, 1993), p. 19; English translation as "Sauf le nom" by John P. Leavey, Jr, in *On the Name*, ed. Thomas Dutoit (Stanford: Stanford University Press, 1995), p. 35.

61. Heidegger, "Nietzsche's Word 'God is Dead,'" p. 236 / 100.

62. Derrida, "The Animal That Therefore I Am," p. 271 / 389.

63. Derrida, "And Say the Animal Responded," p. 137. See also *Of Grammatology*, p. 347 / 244.

64. For more on the importance of tears, see Jacques Derrida, "Donner la mort," in *L'Éthique du don: Jacques Derrida et la pensée du don* (Paris: Transition, 1992), pp. 57–58; English translation by David Wills as *The Gift of Death* (Chicago: University of Chicago Press, 1995), p. 55.

65. For the same comment about Nietzsche crying over the horses, see Derrida, "The Animal That Therefore I Am," p. 286 / 403.

66. Heidegger too recalls this passage in discussing the essence of animality. See Martin Heidegger, *Gesamtausgabe*, vol. 29 / 30, *Die Grundbegriffe der Metaphysik: Welt—Endlichkeit—Einsamkeit* (Frankfurt am Main: Klostermann, 1983), p. 396; English translation by William McNeill and Nicholas Walker as *The Fundamental Concepts of Metaphysics: World, Finitude, Solitude* (Bloomington: Indiana University Press, 1995), p. 273.

4. Eschatology and Positivism

1. Martin Heidegger, "Was ist Metaphysik?" in *Wegmarken* (Frankfurt am Main: Klostermann, 1967), p. 112; English translation by David Farrell Krell, as "What Is Metaphysics?" in *Pathmarks*, ed. William McNeill (Cambridge: Cambridge University Press, 1998), p. 89. Hereafter all essays in *Pathmarks* will be cited as WM with reference first to the German, then to the English translation.

2. Concerning the overcoming of metaphysics in Heidegger, I think that what Heidegger says late in his career, in the 1964 "Time and Being," must not mislead us. Although Heidegger seems to repudiate the intention of overcoming metaphysics, he endorses a thinking that overcomes the obstacles that tend to make a saying of being, without regard for metaphysics, inadequate. More importantly, when he speaks of leaving metaphysics alone, he uses the verb *überlassen*, which also suggests a kind of superengagement with metaphysics. See Martin Heidegger, *Zur Sache des Denkens* (Tübingen: Niemeyer, 1969), p. 25; English translation by Joan Stambaugh as *On Time and Being* (New York: Harper Colophon, 1972), p. 24.

3. One finds a similar critique of the concept of *Erlebnis* in Hans-Georg Gadamer's *Wahrheit und Methode* (Tübingen: Mohr, 1975); English translation by Joel Weinsheimer and Donald G. Marshall as *Truth and Method*, 2d rev. ed. (New York: Continuum, 1989). Gadamer claims that the concept of *Erlebnis* consists in the immediacy of self-consciousness and in an immediacy that yields a content (*das Erlebte*). His critique is that *Erlebnis* is unity and interiority, whereas life itself is self-diremption (*Selbstbehauptung*, "self-differentiation"). Here he takes his inspiration from Hegel, "the speculative import of the concept of life" (*Truth and Method*, pp. 237 / 250–51). Thus, because of the idea of self-diremption, Gadamer stresses the idea of judgment, *Urteil* in German, which literally means "original partitioning." Nevertheless, despite Gadamer's emphasis on *Ur-teil*, I think, with Foucault, that life is not expressed in a judgment, which still relies on unity or synthesis, but in the infinitive of a verb, which can be infinitely divided without unity. It is an expression of the indefinite, a universal singularity. On this idea of the verb, see Gilles Deleuze, *Logique du sens* (Paris: Minuit, 1969), p. 11; English translation by Mark Lester, with Charles Stivale, ed. Constantin

V. Boundas, as *Logic of Sense* (New York: Columbia University Press, 1990), p. 3. See also Foucault's review of Deleuze's *Logic of Sense* and *Difference and Repetition*, "Theatrum Philosophicum," in *Dits et écrits I, 1954–1975* (Paris: Gallimard, 2001), pp. 950–51; English translation in Michel Foucault, *Language, Counter-Memory, Practice*, ed. Donald F. Bouchard (New York: Cornell University Press, 1977), pp. 173–75. If there is a concept in Foucault, it would be an infinitive, like *représenter, classer, parler, échanger, surveiller et punir*, or, finally, *penser*. It is important to recall that Deleuze says that a statement (*un énoncé*) in Foucault—and a statement is the true equivalent to the concept in Foucault—is a "curve" (*un courbe*) (Gilles Deleuze, *Foucault* [Paris: Minuit, 1986], p. 87; English translation by Seán Hand as *Foucault* [Minneapolis: University of Minnesota Press, 1988], p. 80).

4. See Michel Foucault, "Vie: Experience et science," in *Dits et écrits, IV* (Paris: Gallimard, 1994): 763–76; English translation by Robert Hurley as "Life: Experience and Science," in *Essential Works of Michel Foucault: Aesthetics, Method, and Epistemology*, vol. 2, ed. James D. Faubion (New York: The New Press, 1998), pp. 465–78. This project could be completed, it seems, only by a reading of Merleau-Ponty's "L'Homme et l'adversité," in *Signes* (Paris: Gallimard, 1960), pp. 284–308, esp. pp. 299 and 306; English translation by Richard C. McCleary as *Signs* (Evanston, Ill.: Northwestern University Press, 1964), pp. 224–243, esp. pp. 235 and 241. I intend to pursue this question of the *mélange* in Merleau-Ponty in another book project, *Merleau-Ponty and the Political*.

5. Edmund Husserl, Hua III.1: *Ideen zu einer reinen Phänomenologie und phänomenologischen Philosophie*, bk. 1, ed. Karl Schuhmann (The Hague: Martinus Nijhoff, 1976); English translation by F. Kersten as *Ideas pertaining to a Pure Phenomenology and to a Phenomenological Philosophy* (The Hague: Martinus Nijhoff, 1982). See also Edmund Husserl, *Idées directrices pour une phénoménologie*, trans. Paul Ricœur (Paris: Gallimard, 1950).

6. Here I am relying on Husserl's later revision of the passage: "copy D." See Kersten's English translation, p. 73.

7. This solution to the transcendental problem, a solution that defines "psychologism," is circular because it takes something existing in the world, the psyche, which has the ontological sense of something existing in the world, *Vorhandenheit*, and tries to make this something present account for all things present.

8. Edmund Husserl, Hua IX: *Phänomenologische Psychologie* (The Hague: Martinus Nijhoff, 1962), p. 292; English translation by Richard E. Palmer in *The Essential Husserl*, ed. Donn Welton (Bloomington: Indiana University Press, 1999), p. 331.

9. *The Essential Husserl*, p. 294 / 332.

10. Maurice Merleau-Ponty, *Phénoménologie de la perception* (Paris: Gallimard, 1945); English translation by Colin Smith, revised by Forrest Wil-

liams, as *Phenomenology of Perception* (Atlantic Highlands, N.J.: The Humanities Press, 1981). Hereafter PhP, with reference first to the French, then to the English.

11. Thomas Busch also cites this passage in "Maurice Merleau-Ponty: Alterity and Dialogue," in *Circulating Being: Essays in Late Existentialism* (New York: Fordham University Press, 1999), p. 83. Busch's interpretation of Merleau-Ponty's concept of ambiguity and equivocity is based on the idea of dialogue, an idea very different from a battle.

12. Derrida makes a similar comment, stressing form and content, in the Introduction to *Voice and Phenomenon*: presence has always been and will always be, to infinity, the form in which—we can say this apodictically—the infinite diversity of content will be produced. The opposition—which inaugurates metaphysics—between form and matter finds in the concrete ideality of the living present its ultimate and radical justification (VP 5 / 6).

13. Spacing implies what Derrida calls "archi-writing" and thus vision: "when I see myself writing and when I signify by gestures, the proximity of hearing myself speak is broken" (VP 90 / 80). Thus here we could speak of a story of the eye, which would allow for a further comparison with Foucault.

14. Gilles Deleuze, *Différence et répétition* (Paris: Presses Universitaires de France, 1968), p. 179; English translation by Paul Patton as *Difference and Repetition* (New York: Columbia University Press, 1994), p. 137. Deleuze, *The Logic of Sense*, pp. 119 and 124 / 97 and 102.

15. See Deleuze, *Difference and Repetition*, p. 177 / 135; also Deleuze, *The Logic of Sense*, p. 54 / 39.

16. Deleuze, *The Logic of Sense*, p. 120 / 99, my emphasis.

17. I have coined a word for this idea of "a promise that demands to be done over again and again": *refinition*. See the preface to my *Derrida and Husserl*, also *Thinking through French Philosophy*.

18. This description is based largely on Michel Foucault, *Ceci n'est pas une pipe* (Paris: Fata Morgana, 1973); English translation by James Harnes as *This Is Not a Pipe* (Berkeley: University of California Press, 1983). See also Deleuze, *Foucault*, p. 119 / 112.

19. I am extrapolating from what Foucault has said in *L'Ordre du discours* (Paris: Gallimard, 1971), p. 72; English translation by A. M. Sheridan as "The Discourse on Language," appendix to *The Archeology of Knowledge* (New York: Pantheon, 1972), p. 234.

20. If we were to pursue farther this difference between Foucault and Derrida, we would have to investigate the concept of multiplicity.

21. Foucault, "The Discourse on Language," p. 72 / 234.

5. *Un écart infime* (Part I)

1. Michel Foucault, "Vie: Experience et science," in *Dits et écrits, IV* (Paris: Gallimard, 1994), pp. 763–76; English translation by Robert Hurley

as "Life: Experience and Science," in *Essential Works of Michel Foucault: Aesthetics, Method, and Epistemology*, vol. 2, ed. James D. Faubion (New York: The New Press, 1998), pp. 465–78. We shall refer to this text with the abbreviation VES, with reference first to the French, then to the English translation.

2. Here we must think of Deleuze, too, since the title of his last text, in 1995, is "Immanence: Une vie."

3. This project could complete itself, it seems, only by a reading of Merleau-Ponty's "L'Homme et l'aversité," in *Signes* (Paris: Gallimard, 1960), pp. 299 and 306; English translation by Richard C. McCleary as *Signs* (Evanston, Ill.: Northwestern University Press, 1964), pp. 235 and 241. Hereafter cited by the abbreviation S, with reference first to the French edition, then to the English translation.

4. Michel Foucault, *Ceci n'est pas une pipe* (Montpelier: Fata Morgana, 1973); English translation by James Harnes as *This Is Not a Pipe* (Berkeley: University of California Press, 1983). Hereafter cited as CP, with reference first to the French, then to the English. We can say that *Ceci n'est pas une pipe* is roughly contemporaneous with *Words and Things* because a first version of it exists from 1968. See *Dits et écrits I, 1954–1975* (Paris: Gallimard, 2001), pp. 663–78.

5. Gilles Deleuze, *Foucault* (Paris: Minuit, 1986), p. 89; English translation by Seán Hand as *Foucault* (Minneapolis: University of Minnesota Press, 1988), p. 83.

6. See also Maurice Merleau-Ponty, *La Structure du comportement* (Paris: Presses Universitaires de France, 1942), p. 232; English translation by Alden L. Fisher as *The Structure of Behavior* (Pittsburgh: Duquesne University Press, 1983), p. 215. Merleau-Ponty appropriates the idea of a *mélange* from Descartes's Sixth Meditation. See Maurice Merleau-Ponty, *L'Union de l'âme et du corps chez Malebranche, Biran et Bergson* (Paris: Vrin, 1978), p. 13; English translation by Paul B. Milan as *The Incarnate Subject* (Amherst, N.Y.: Humanity Books, 2001), p. 33; see also p. 81/89 for the connection of *mélange* to Bergson.

7. He also speaks of *une histoire vécu*; see PhP, p. 512/449.

8. In a remarkable essay, Rudolf Boehm has tracked the ambiguity concerning the concepts of immanence and transcendence in Husserl. He says that in the *Logical Investigations* immanence means strictly lived-experience, without any transcendent object, and, in this regard, immanence was not ambiguous for Husserl. But in the *Idea of Phenomenology*, Husserl (anticipating the concept of the noema) includes the intentional object in immanence, since he thought it to be given in an adequate intuition. From 1907 on (the time of the lectures that make up *The Idea of Phenomenology*), the Husserlian concept of *Erlebnis* is ambiguous, a mixture. See Rudolf Boehm, "Les Ambiguities des concepts husserliens d'"immanence' et de 'transcendence,'" in *Revue philosophique de la France et de l'étranger* 84 (1959): 481–526.

9. Thomas Busch also cites this passage in "Maurice Merleau-Ponty: Alterity and Dialogue," in *Circulating Being: Essays in Late Existentialism* (New York: Fordham University Press, 1999), p. 83. Busch's interpretation of Merleau-Ponty's concept of ambiguity and equivocity is based on the idea of dialogue, an idea very different from a battle.

10. All the citations in the next two paragraphs are taken from the section called "The Empirical and the Transcendental" of chapter 9 of *Words and Things*, 329–33 / 318–22.

11. Michel Foucault, *L'Histoire de la folie à l'âge classique* (Paris: Gallimard, 1972), p. 310. The other citations come from the final pages of the chapter called "La Transcendence du délire," pp. 309–18.

12. Ibid., p. 240. See also p. 265.

13. Ibid., p. 239.

14. Deleuze, *Foucault*, p. 119 / 112. See also Michel Foucault, "Préface à la transgression," in *Dits et écrits I*, pp. 261–78, esp. p. 266; English translation by Donald F. Bourchard as "Preface to Transgression," in *Essential Works of Foucault: Aesthetics, Method, and Epistemology*, pp. 69–87, esp. p. 74. On this page, Foucault cites Kant's early essay on negative magnitudes; here, on the basis of how negative numbers function in relation to positive numbers, Kant stresses that real conflict occurs only between two positive forces. See Immanuel Kant, *Theoretical Philosophy 1755–1770*, ed. David Walford (Cambridge: Cambridge University Press, 1992), pp. 206–41, esp. p. 215.

15. Michel Foucault, "La Pensée du dehors," in *Dits et écrits, I*, pp. 546–67; English translation by Brian Massumi as "The Thought from Outside," in *Foucault / Blanchot* (New York: Zone Books, 1997), pp. 7–60. At the beginning of the third section, "Reflection, Fiction," Foucault says that the thought of the outside is not lived-experience.

16. In *Foucault*, Deleuze says that "Truly, one thing haunts Foucault and that is thought, 'what does thinking mean, what calls for thinking,' the question shot by Heidegger, and taken up by Foucault, the arrow par excellence" (p. 124 / 116; my translation, my emphasis). Also in *Ceci n'est pas une pipe*, Foucault describes *la bataille*, "the battle," as "arrows shot at the target of the adversary" (CP 30 / 26).

17. See again Foucault's "Preface to Transgression," in which he cites Kant's pre-Critical essay on negative magnitudes. It seems probable that Foucault's idea of untying and analyzing comes from this essay and Kant's 1764 *Inquiry*. See Immanuel Kant, *Kants Werke*, vol. 2 (Berlin: Georg Reimer, 1912); English translations can be found in *Theoretical Philosophy 1755–1770* (Cambridge: Cambridge University Press, 1992). It is perhaps accidental (or perhaps due to the influence of Hyppolite) that Foucault's methodology overlaps with that of Bergson: differentiating or dissociating a mixture.

18. We should, of course. recall Eugen Fink's famous *Kantstudien* essay "Die Phänomenologische Philosophie E. Husserl in der Gegenwärtigen

Kritik," originally published in *Kantstudien* 38, nos. 3/4 (Berlin, 1933); collected in Fink, *Studien zur Phänomenologie* (The Hague: Martinus Nijhoff, 1966); English translation as "The Phenomenological Philosophy of Edmund Husserl and Contemporary Criticism," in *The Phenomenology of Husserl*, ed. R. O. Elveton (Chicago: Quadrangle Books, 1970), pp. 73–147. For the French translation, see Eugen Fink, *De la phénoménologie*, trans. Didier Franck (Paris: Minuit, 1974).

19. Another way of describing this difference would be to say that phenomenology forgot the difference between beings and being.

20. Maurice Merleau-Ponty, *Le Visible et l'invisible* (Paris: Gallimard, 1964), p. 328; English translation by Alphonso Lingis as *The Visible and the Invisible* (Evanston, Ill.: Northwestern University Press, 1968), pp. 274–75. Hereafter cited as VI, with reference first to the original French, then to the English translation. See also Myriam Revault d'Allonnes, *Merleau-Ponty: La Chair du politique* (Paris: Michalon, 2001), p. 96, where she claims that, for Merleau-Ponty, Marx opens up an "equivocal power."

21. Busch, "Maurice Merleau-Ponty," charts the transition from the early Merleau-Ponty of *Phenomenology of Perception* to the later ontology of *The Visible and the Invisible*; he stresses quite correctly that the notion of institution replaces that of *vécu*.

22. Maurice Merleau-Ponty, *L'Œil et l'esprit* (Paris: Gallimard, 1964), p. 87; English translation by Galen John and Michael B. Smith as "Eye and Mind," in *The Merleau-Ponty Aesthetics Reader* (Evanston, Ill.: Northwestern University Press, 1993), p. 148. Hereafter cited as OE with reference first to the original French, then to the English translation. For more on Merleau-Ponty's relation to Klee, see Galen Johnson, "Ontology and Painting: 'Eye and Mind,'" in *The Merleau-Ponty Aesthetics Reader*, pp. 35–55, esp. pp. 39–44.

23. In fact, Merleau-Ponty associates this immanence with Descartes. See Maurice Merleau-Ponty, *Notes de cours, 1959–1961* (Paris: Gallimard, 1996), p. 180nA.

24. Here Merleau-Ponty also alludes to this comment by Klee.

25. See Michel Foucault, *Dits et écrits, I, 1954–1975* (Paris: Gallimard, 2001), p. 572. See also Gary Shapiro, *Archeologies of Vision* (Chicago: University of Chicago Press, 2003), pp. 276–77.

26. See Maurice Merleau-Ponty, "Annexe: L'Homme et l'adversité," in *Parcours Deux, 1951–1961* (Lagrasse: Verdier, 2000), p. 340, where he distinguishes between ambivalence and ambiguity. Ambiguity consists in contraries participating in one same being, while ambivalence is to have two alternating images of one same being.

27. See Thomas R. Flynn, *Sartre, Foucault, and Historical Reason: Towards an Existentialist Theory of History*, vol. 1 (Chicago: University of Chicago Press, 1997). Stressing ambivalence and ambiguity in Sartre's reflections on history, Flynn almost never speaks of Merleau-Ponty.

28. Maurice Merleau-Ponty, *Les Aventures de la dialectique* (Paris: Gallimard, 1955), p. 316; English translation by Joseph Bien as *Adventures of the Dialectic* (Evanston, Ill.: Northwestern University Press, 1973), p. 228.

29. See also Merleau-Ponty, *The Structure of Behavior*, p. 212 / 197.

30. See also Merleau-Ponty, *Signs*, p. 29 / 13, and *Notes de cours, 1959– 1961*, where, in "L'Ontolgie cartesienne et l'ontologie d'aujourd'hui," he repeatedly claims that Descartes's thought is ambiguous, *un mélange* of the body and the soul (pp. 188, 235, 242, 267, 268). See also Renaud Barbaras, *Le Tournant de l'expérience* (Paris: Vrin, 1998), p. 53; here Barbaras speaks of an originary *mélange* in Merleau-Ponty.

31. The English translation renders *implication* as "involvement." See also VI 187 / 142, where Merleau-Ponty speaks of a "Sentient in general," a "Sensibility in general," and "this generality."

32. Daniel N. Robinson, *Significant Contributions to the History of Psychology, 1750–1920* (Washington, D.C.: University Publications of America, 1978), p. 10. This volume contains a reprint of F. Gold's 1827 English translation of Bichat's *Investigations*. Xavier Bichat was a French anatomist and biologist, who lived from 1771 to 1802. He published four books: *Traité des membranes* (1799); *Recherches physiologiques sur la vie et la mort* (1800); *Anatomie générale appliquée à la médecine* (1801); *Anatomie descriptive* (1801). For a short biographical sketch, see Xavier Bichat, *Recherches physiologiques sur la vie et la mort (première partie) et autres textes* (Paris: Flammarion, 1994), pp. 389–91. It is perhaps significant that this biographical sketch speculates that Bichat died from something he contracted after opening up some corpses. There is an English translation of the *Recherches*: *Physiological Researches on Life and Death* (Washington, D.C.: University Publications of America, 1978).

33. Michel Foucault, *Naissance de la clinique* (Paris: Presses Universitaires de France, 1963), p. 147; English translation by A. M. Sheridan Smith as *The Birth of the Clinic: An Archaeology of Medical Perception* (New York: Vintage, 1973), p. 144. Hereafter cited as NC with reference first to the French, then to the English translation.

34. Although written before having read this book (in the spring and summer of 2003), this chapter intersects completely with Gary Shapiro's magnificent *Archeologies of Vision: Foucault and Nietzsche on Seeing and Saying* (Chicago: University of Chicago Press, 2003).

35. See also Michel Foucault, *Surveiller et punir* (Paris: Gallimard, 1975), p. 224; English translation by Alan Sheridan as *Discipline and Punish* (New York: Vintage, 1978), p. 191.

36. Ibid., p. 292 / 252.

6. *Un écart infime* (Part II)

1. Rudolf Boehm, "Les Ambiguities des concepts husserliens d'immanence' et de 'transcendence,'" in *Revue philosophique de la France et de l'étranger* 84 (1959): 481–526.

2. See chapter 5 in this book.

3. MC 332 / 321 . See also S 229 / 235, 306 / 241.

4. For more on Merleau-Ponty's relation to Klee, see Galen Johnson, "Ontology and Painting: 'Eye and Mind,'" in *The Merleau-Ponty Aesthetics Reader*, ed. Galen Johnson (Evanston, Ill.: Northwestern University Press, 1993), pp. 35–55, esp. pp. 39–44.

5. In 1961, it appeared in *Art de France*. In 1964 it appeared as a small book from Gallimard.

6. "L'Ontologie cartesienne et l'ontologie d'aujourd'hui"; see Maurice Merleau-Ponty, *Notes de cours, 1959–1961* (Paris: Gallimard, 1996). Hereafter cited as NdC 1959–61. All translations are my own.

7. See also NdC 1959–61 264, where Merleau-Ponty says that Descartes is the most difficult of authors because he is the most radically ambiguous; Descartes, Merleau-Ponty says, has the most latent content. Merleau-Ponty makes the same comments about Descartes in the first nature lectures course. See Maurice Merleau-Ponty, *La Nature: Notes de cours du Collège de France*, ed. Dominque Seglard (Paris: Seuil, 1995), pp. 36–37; English translation by Robert Vallier as *Nature: Course Notes from the Collège de France* (Evanston, Ill.: Northwestern University Press, 2003), pp. 17–18. In the nature lectures, Merleau-Ponty also says that nature is a mixture (p. 164 /121).

8. Maurice Merleau-Ponty, *L'Union de l'âme et du corps chez Malebranche, Biran et Bergson* (Paris: Vrin, 1978), p.13; English translation by Paul B. Milan as *The Incarnate Subject: Malebranche, Biran, and Bergson on the Union of Body and Soul* (Amherst, N. Y.: Humanity Books, 2001), p. 33. In this passage, Merleau-Ponty is quoting Descartes's Sixth Meditation. The quote can be found on p. 192 of the Haldane and Ross translation of the *Meditations* (*The Philosophical Writings of Descartes* [Cambridge: Cambridge University Press, 1973]). See also *Meditationes de prima philosophia, Méditations métaphysiques*, Latin text with French translation by Duc de Luynes (Paris: Vrin, 1978), p. 81, l. 13: in the Latin, *permixtione*; *mélange* in the Duc's French translation.

9. See Galen Johnson's introduction to "Eye and Mind" in *The Merleau-Ponty Aesthetics Reader*, pp. 35–55. Johnson claims that Merleau-Ponty's philosophy of the flesh, the philosophy opposed to great rationalism, is not an ontological monism, not "a metaphysics of substance and sameness, a monism of the One" (p. 49). The concept of sameness that I am attributing to Merleau-Ponty, his mixture, is not a reductive identity, as I am trying to show through the three conceptual schemes. It is the sameness of identity and difference. Sartre's philosophy, according to Merleau-Ponty, is an ontological monism.

10. Maurice Merleau-Ponty, *La Structure du comportement* (Paris: Presses Universitaires de France, 1990); English translation by Alden L. Fisher as *The Structure of Behavior* (Pittsburgh: Duquesne University Press, 1983). Hereafter cited as SC, with reference first to the French, then to the English translation.

11. We are justified in returning to this work, which is nearly twenty years earlier than "Eye and Mind," because in the course "Descartes's Ontology and Today's Ontology" Merleau-Ponty makes use of the figure-ground formula of the Gestalt when he criticizes Descartes's theory of vision. We shall return to this critique. See NdC 1959–61, 229.

12. Deleuze begins his examination of Spinoza by referring to this passage from Merleau-Ponty. See Gilles Deleuze, *Spinoza et le problème de l'expression* (Paris: Minuit, 1968), p. 22; English translation by Martin Joughin as *Expressionism in Philosophy: Spinoza* (New York: Zone Books, 1990), p. 28. It is also clear that this distinction between the positive infinite and the indefinite maps onto Hegel's distinction between the good infinite and the bad infinite, but Merleau-Ponty never mentions that.

13. See Renaud Barbaras, who clearly sees the connection between Merleau-Ponty and Leibniz; *The Being of the Phenomenon* (Bloomington: Indiana University Press, 2004), pp. 229–34.

14. While all commentators have noted the relation of "Eye and Mind" to Descartes, no one, so far as I know, has presented its central thesis as being about the heritage of large rationalism. In particular, see: Hugh J. Silverman, "Cézanne's Mirror Stage," in *The Merleau-Ponty Aesthetics Reader*, pp. 262–77, esp. p. 265; Véronique Fóti, "The Dimension of Color," in *The Merleau-Ponty Aesthetics Reader*, pp. 293–308, esp. pp. 296–97; and François Cavallier, *Premières leçons sur L'Œil et l'esprit de M. Merleau-Ponty* (Paris: Presses Universitaires de France, 1998), pp. 38–46. None of the commentators systematize Merleau-Ponty's analysis of Descartes's *Optics*. Galen Johnson's introduction to "Eye and Mind" in *The Merleau-Ponty Aesthetics Reader*, pp. 35–55, while excellent in many regards, does not mention Descartes.

15. In *Words and Things*, Foucault describes the exact relation to the infinite that Merleau-Ponty here is describing. Foucault says that the relation to the infinite in the classical epoch (Cartesianism) was a "negative relation." See MC 327 / 316.

16. This discussion should be compared to the one found in the nature lectures (cf. *Nature*, pp. 169–76 / 125–31).

17. See again: Silverman, "Cézanne's Mirror Stage," in *The Merleau-Ponty Aesthetics Reader*, p. 265; Fóti, "The Dimension of Color," in *The Merleau-Ponty Aesthetics Reader*, pp. 296–97; and Cavallier, *Premières leçons sur L'Œil et l'esprit de M. Merleau-Ponty*, pp. 38–46. Some commentators recognize that for Merleau-Ponty vision in Descartes is conceived as thought (Silverman), while others stress the model of touch (Fóti). Cavallier notes that Merleau-Ponty discusses Descartes's different "models" for vision (touch and thought) but does not see the different models as being related (p. 38).

18. See also Mauro Carbone, *The Thinking of the Sensible* (Evanston, Ill.: Northwestern University Press, 2004), esp. pp. 45–47; here Carbone stresses the literal sense of *concept* as "to grasp."

19. In *Le Visible et l'invisible*, Merleau-Ponty says *on se tromperait*, "one would be mistaken."

20. It is well known that Descartes tried consciously to break with the Scholastic tradition and used the *Summa Philosophica Quadripartita* of Eustache de Sancto Paulo as his guide to Scholastic philosophy. An intentional species for the Scholastics, according to Eustache, is a mental image, but not a copy of an individual thing. It is an exemplar or species, an *eidos*, the Greek equivalent of species. Apparently, the discussion of ideas throughout the Scholastic period always referred to painters, or more generally, artists. The model would be the exemplar or idea or intentional species, while the painting would be the image, the particular. Referring back to the *Timaeus*, this discussion conceived God as an artificer. See Roger Ariew, *Descartes and the Last Scholastics* (Ithaca, N.Y.: Cornell University Press, 1999), pp. 64–69. What is important for our purposes is that the concept of intentional species implies some sort of resemblance relation.

21. For more on Merleau-Ponty and the geometral, see Jacques Lacan, *Les Quartes Concepts fondamentaux de la psychanalyse* (Paris: Seuil, 1973), chap. 2; English translation by Alan Sheridan as *The Four Fundamental Concepts of Psycho-Analysis* (New York: Norton, 1978), chap. 2, "Of the Subject of Certainty."

22. See Deleuze, *Expressionism in Philosophy*, pp. 38 / 46–47.

23. In his book on Foucault, Deleuze cites these final pages of *The Visible and the Invisible* (VI 201–2 / 153–54). See Gilles Deleuze, *Foucault* (Paris: Minuit, 1986), p. 119n39; English translation by Seán Hand as *Foucault* (Minneapolis: University of Minnesota Press, 1988), p. 149n38.

24. For more on Klee, Merleau-Ponty, and auto-figuration, see Stephen Watson, "On the Withdrawal of the Beautiful: Adorno and Merleau-Ponty's Readings of Klee," in *Chiasmi International* 5 (2003): 201–21. See also Galen Johnson, "Thinking in Color: Merleau-Ponty and Klee," in *Merleau-Ponty: Difference, Materiality, Painting*, ed. Veronique Fóti (Atlantic Highlands, N.J.: Humanities Press, 1996).

25. "L'Entrelacs—le chiasme" is, of course, the title of *The Visible and the Invisible*'s fourth chapter.

26. Merleau-Ponty, in fact, says that art, once it has awakened, gives vision new powers; these powers would have to define the painter's vision.

27. For the same quote, see also NdC 1959–61 206.

28. See Françoise Dastur's "La Pensée du dedans," in *Chair et langage: Essais sur Merleau-Ponty* (Paris: Encre Marine, 2001), esp. pp. 125–26, where she compares, but not *a contrario*, Merleau-Ponty's "thought of the inside" to Foucault's "thought of the outside."

29. See also my "The Legacy of Husserl's 'The Origin of Geometry': The Limits of Phenomenology in Merleau-Ponty and Derrida," in *Thinking through French Philosophy* (Bloomington: Indiana University Press, 2033), pp. 62–79. At the time of writing that essay (1999), I was not aware of the difference in emphasis that this imminence makes. See LT 238–40.

30. Maurice Merleau-Ponty, *Les Relations avec autrui chez l'enfant* (Paris: Centre de Documentation Universitaire, 1960), p. 55; English translation by William Cobb as "The Child's Relation with Others," in *The Primacy of Perception* (Evanston, Ill: Northwestern University Press, 1964), p. 135. The lectures date from 1949–51. In reference to the difference between the imaginary and the symbolic, see Gilles Deleuze, "A quoi reconnaît-on le structuralisme?" in *L'Île déserte et autres textes* (Paris: Minuit, 2002), pp. 238–69; English translation by Melissa McMahon and Charles J. Stivale as "How Do We Recognize Structuralism?" in *Desert Islands and Other Texts* (New York: Semiotext(e), 2004), pp. 170–92.

31. For more on blindness in Merleau-Ponty, see Galen Johnson, "The Retrieval of the Beautiful," unpublished manuscript, 2004. I completed all three parts of this trilogy before reading Johnson's essay, which he was kind enough to share with me.

32. For more on verticality and vision, see Erwin W. Strauss, "The Upright Posture," in *Psychiatric Quarterly* 26, no. 4 (October 1952): 529–61, esp. p. 546.

33. For more on the question of man in both Merleau-Ponty and Foucault, see also Etienne Bimbinet, *Nature et Humanité: Le Probléme anthropologique dans l'œuvre de Merleau-Ponty* (Paris: Vrin, 2004), esp. pp. 312–13.

7. *Noli me tangere*

1. See John Russon, *Human Experience: Philosophy, Neurosis, and the Elements of Everyday Life* (Albany: State University of New York Press, 2003). See my review of Russon's excellent book, "'Noli me tangere': Reflections on Vision Starting from John Russon's *Human Experience*," in *Continental Philosophy Review* (forthcoming).

2. See Françoise Dastur, "Monde, Chair, Vision," in *Chair et langage: Essais sur Merleau-Ponty* (Paris: Encre Marine, 2001), pp. 97–99.

3. For more on verticality and distance, see Erwin W. Strauss, "The Upright Posture," in *Psychiatric Quarterly* 26, no. 4 (October 1952): 529–61, esp. p. 546.

4. See LT 120.

8. *Un écart infime* (Part III)

1. Michel Foucault, "La Pensée du dehors,' in *Dits et écrits, I, 1954–1969* (Paris: Gallimard, 1994), p. 552; English translation by Brian Massumi as "The Thought from Outside," in *Foucault/Blanchot* (New York: Zone Books, 1990), p. 24, my emphasis.

2. This a-perspectivism, a term I am adapting from MdA 48 / 44–45, does not eliminate the idea that Foucault appropriates Nietzsche's perspectivism (see Michel Foucault, "Nietzsche, la généaologie, l'histoire," in *Dits et écrits I, 1954–1975* [Paris: Gallimard, 2001], p. 1015; English translation as

"Nietzsche, Genealogy, History" by Donald F. Bourchard and Sherry Simon, in *Essential Works of Foucault, 1954–1984*, vol. 2: *Aesthetics, Method, Epistemology* [New York: The New Press, 1998]). Rather, it makes the concept more precise. Foucault's perspectivism has no object at the center around which the perspectives would be organized. The perspectives therefore amount to a series, a series of errors. In this regard, it is also important to recall the idea of anamorphosis. See Jacques Lacan, *Les Quartes Concepts fondamentaux de la psychanalyse* (Paris: Seuil, 1973), pp. 92–104; English translation by Alan Sheridan as *The Four Fundamental Concepts of Psycho-Analysis* (New York: Norton, 1981), pp. 79–90. A famous painting in which one finds anamorphosis is Holbein's *The Ambassadors*. See also Gilles Deleuze, *Le Pli: Leibniz et le Baroque* (Paris: Minuit, 1988), p. 27; English translation by Tom Conley as *The Fold: Leibniz and the Baroque* (Minneapolis: University of Minnesota Press, 1993), p. 20. Here Deleuze speaks of a very specific form of perspectivism, which still revolves around the idea of an object $= X$.

3. Michel Foucault, "Préface à la transgression," in *Dits et écrits*, *I*, p. 246; English translation by Donald F. Bourchard as "A Preface to Transgression,' in *Essential Works of Foucault*, 2:82 (my emphasis): "Death is not for the eye the line that is always receding from the horizon, but in its very location [*emplacement*], in the hollow of all its possible gazes, [it is] the limit that it transgresses, making it surge as the *absolute limit* in the ecstatic movement which allows it to leap to the other side. The upturned eye discovers the connection of language to death in the moment when it figures the play of the limit and of being. Perhaps the explanation of the prestige of the eye is that it founds the possibility of giving a language to this play."

4. Cf. NC 144 / 142. Indeed, we must characterize the invisible in Foucault as "the visible invisible." See also NC 174 / 170: "Language and death have operated at every level of this experience, and in accordance with its whole density, only to offer at last to scientific perception what, for it, had remained for so long the *visible invisible*—the forbidden, imminent secret: the knowledge of the individual."

5. Michel Foucault, "Vie: Experience et science," in *Dits et écrits*, *IV* (Paris: Gallimard, 1994), p. 776; the English translation is by Robert Hurley as "Life: Experience and Science," in *Essential Works of Michel Foucault*, 2:477.

6. Gilles Deleuze, *Foucault* (Paris: Minuit, 1986), p. 124; English translation by Seán Hand as *Foucault* (Minneapolis: University of Minnesota Press, 1988), p. 116, my emphasis.

7. The following texts have been consulted in the writing of this essay: Véronique Fóti, *Vision's Invisibles: Philosophical Investigations* (Albany: The State University of New York Press, 2003); Jonathan Brown, *Images and Ideas in Seventeenth-Century Spanish Painting* (Princeton: Princeton University

Press, 1978); Svetlana Alpers, "Interpretation without Representation; or, The Viewing of *Las Meninas*," *Representations* 1, no. 1 (February 1983): 31–42; Claude Gandelman, "Foucault as Art Historian," *Hebrew University Studies in Literature and the Arts* 13, no. 2 (autumn 1985): 266–80; Leo Steinberg, "Velasquez' *Las Meninas*," *October* 19 (1981): 45–54; John R. Searle, "*Las Meninas* and the Paradoxes of Pictorial Representation," *Critical Inquiry* 6 (1980): 177–88; Joel Snyder and Ted Cohen, "Critical Response: Reflexions on *Las Meninas: Paradise Lost*," *Critical Inquiry* 7 (1980): 129–47; Joel Synder, "*Las Meninas* and the Mirror of the Prince," *Critical Inquiry* 11 (June 1985): 539–72; John Carvalho, "The Visible and the Invisible: In Merleau-Ponty and Foucault," *International Studies in Philosophy* 25, no. 3 (1993): 35–46; Helen Fielding, "Depth of Embodiment: Spatial and Temporal Bodies in Foucault and Merleau-Ponty," *Philosophy Today* 43 (Spring 1999): 73–85; Stephen Watson, "Merleau-Ponty and Foucault: De-Aestheticization of the Work of Art," *Philosophy Today* 28 (Summer 1984): 148–65.

8. In the English translation, the name *Velasquez* appears on p. 6, but this name is an unreflective translation of the French word *auteur*.

9. Michel Foucault, *La Peinture de Manet* (Paris: Seuil, 2004). The preface to this volume tells us that Foucault presented this lecture in 1971 but that it dates back to 1967. The preface also states that Foucault had a contract with Minuit for a book on Manet in 1967.

10. Michel Foucault, "'Les Suivantes,'" *Mercure de France* 354 (July–August 1965): 367–384.

11. See *Dits et écrits, I*, p. 544.

12. See Michel Foucault, *L'Archéologie du savoir* (Paris: Gallimard, 1969), pp. 168–69; English translation by A. M. Sheridan Smith as *The Archeology of Knowledge* (New York: Pantheon, 1972), p. 128.

13. See NC 166 / 162. The role of the sign in nineteenth-century anatomo-clinical experience consists in throwing on the living body a whole network of anatomo-pathological mappings (*repérages*). Thus removing these signs by artifice—they were put there by artifice, too—removes the map or the measure of the space.

14. Foucault, "Nietzsche, Genealogy, History," p. 1017 / 381, my emphasis.

15. Ibid., p. 1012 / 377. The nonplace would be the battlefield, the place of the war that, for Foucault, defines his concept of historicism. See Michel Foucault, *"Il faut défendre la société," Cours au Collège de France, 1976* (Paris: Gallimard and Seuil, 1997), pp. 153–54; English translation by David Macey as *"Society Must Be Defended," Lectures at the Collège de France, 1975–1976* (New York: Picador, 2003), pp. 172–73.

16. See, e.g., CP 30 / 26.

17. I am insisting on the word *analysis* because of Foucault's own use of it in chapter 9 and because he is stressing the idea of de-composition in the painting, a literal analysis or taking apart.

18. Gilles Deleuze, "A quoi reconnaît-on le structuralisme?" in *L'Île dé-serte et autres textes* (Paris: Minuit, 2002), p. 261; English translation by Me-lissa McMahon as "How Do We Recognize Structuralism?" in *Desert Islands and Other Texts* (New York: Semiotext(e), 2004), p. 186. Hereafter abbrevi-ated as QRS with reference first to the French, then to the English translation.

19. It seems to me that this definition of the painting through reciprocity allows one to avoid all the controversy concerning the painting's perspective and Foucault's analysis of it. The principal figures are gazing out at the spectator-model, who is gazing back at them, turning them into a spectacle.

20. Cf. Michel Foucault, "Préface à la transgression," in *Dits et écrits, I*, p. 236; English translation by Sherry Bourchard in *Essential Writings of Michel Foucault*, 2:73: "The line [*le trait*] that it crosses could be its whole space. The play of the limits and of transgression seems to be ruled over by a sim-ple obstinacy: the transgression crosses and does not stop to begin again to cross a line that, behind it, immediately closes itself in, in a wave that is hardly remembered, withdrawing [*reculant*] again up as far as the horizon."

21. For more on the trait, see MdA and Chapter 3 of the present book. The German translation of *Les Mots et les choses* uses *Tupfer* to translate *trait*, and *steht entfernt* to translate *est en retrait*. See Michel Foucault, *Die Ordnung der Dinge*, trans. Ulrich Köppen (Frankfurt am Main: Suhrkamp, 1974), p. 31. In French, *trait* and *retrait* could be used to translate Heidegger's *Entzie-hung* and *Zug*, as Derrida suggests in "Le Retrait de la métaphore," in *Psyché: Inventions de l'autre* (Paris: Galilée, 1987), p. 86.

22. On the difference between *le regard* and *le coup d'œil*, see NC 122–23 / 121–22. Foucault's use here of *coup d'œil*, where the eye functions as fire in a chemical combustion, anticipates the chemical sense of *metathesis* later in the chapter.

23. Before Foucault turns to the gaze of the painting, he speaks of a num-ber of sidesteps, *esquives*, in which the painting consists. These sidesteps in the painting, such as the ideal location of the spectator, try to cover over the blind spot. Cf. Foucault, *La Peinture de Manet*, p. 23: "We have, if you will, the play of side-stepping, of hiding, of illusion and elision, that Western rep-resentational painting has practiced since the Quatrocentro." See also NC 37 / 38: here Foucault speaks of the opposition between two forms of medi-cine being *esquivé* by the "too visible prestige" of a consequence of the two kinds of medicine.

24. On indifference, see Gilles Deleuze, *Différence et repetition* (Paris: Presses Universitaires de France, 1968), p. 43; English translation by Paul Patton as *Difference and Repetition* (New York: Columbia University Press, 1994), p. 28. It is important that Deleuze says here that difference in indif-ference is "the only extreme, the only moment of presence and precision."

25. With regard to this question, see Gary Shapiro's excellent *Archeologies of Vision: Foucault and Nietzsche on Saying and Seeing* (Chicago: University of Chicago Press, 2003), p. 251.

26. Here we could speak of a tain of the mirror; the tain is the backside of the mirror. This phrase, "the tain of the mirror," comes from Rodolphe Gasché's *The Tain of the Mirror: Derrida and the Philosophy of Reflection* (Cambridge: Harvard University Press, 1986). Gasché, of course, takes the phrase from Derrida; see Jacques Derrida, *La Dissémination* (Paris: Seuil, 1971), pp. 39, 349; English translation by Barbara Johnson as *Dissemination* (Chicago: University of Chicago Press, 1981), pp. 33, 314. For more on the relation between Derrida and Foucault, see my *Thinking through French Philosophy: The Being of the Question* (Bloomington: Indiana University Press, 2003). The central thesis of my book is that there is a point of thinking from which the singular thought of Derrida, Deleuze, and Foucault diffracts. This point is the idea of repetition, or *re-trait*, or re-flection: the tain of the mirror. See also Chapter 4 of this book.

27. In the original version of the chapter in *Mercure de France*, a paragraph describes the paintings on the wall; Foucault elides this paragraph in 1966. The reason for this elision is obvious: to make the transition to the mirror at the back of the painting more direct. In the 1966 *Words and Things* version, we go directly to the mirror on the back wall.

28. The 1965 version explains this idea more fully. There Foucault says "the mirrors [in Dutch genre paintings] were like another gaze, which could seize things from behind or on the bias . . . the mirrors made the visibility of things flip over, in order to restore visibility to the order of the picture."

29. The German translation renders *se confondent* as *vermischen sich*.

30. By stressing the uncertainties and the "maybe," which is not hard to do, since Foucault himself repeats them, one can avoid the controversy around the "facts" of the mirror in the Velásquez painting. Foucault himself says that the mirror is pretending and not honest, and therefore it might be showing something other than the royal couple. In light of Foucault's admission of uncertainty, it is hard to understand Joel Snyder's insistence, in "*Las Meninas* and the Mirror of the Prince," that Foucault's analysis depends upon getting the "facts" of the painting right. See note 7, above.

31. See also Lacan, *The Four Fundamental Concepts of Psycho-Analysis*, p. 88 / 75. Here Lacan says, "The subject [in the dream] does not see where it [*ça*: the Id] is going, he follows [*suit*]." *Suit*, of course, is a homonym of *suis*, as in *je suis*, the first-person singular of the French verb *être*, "to be," which implies that what is at issue here is the question of being, the being of the "cogito."

32. My thanks to R. Matilde Mésavage, who stressed to me the feminine aspect of this term *les suivantes*. While this idea goes beyond anything that Foucault has done (so far as I know), it is possible to think that being feminine might be a condition of being a follower, of engaging in this practice that is not sovereign.

33. For the history of the title of the painting, see Shapiro, *Archeologies of Vision*, pp. 245–46.

34. Being followers of a directive is perhaps what Heidegger has in mind with his use of the term *Besinnung*.

35. Or the feeling of madness. See Michel Foucault, *Histoire de la folie à l'âge classique* (Paris: Gallimard, 1972), p. 309.

36. See Michel Foucault, *Surveiller et punir* (Paris: Gallimard, 1975), p. 52; English translation by Alan Sheridan as *Discipline and Punish* (New York: Vintage, 1995 [1977]), p. 41.; and NC 61 / 60.

37. Michel Foucault, "Preface to Transgression, *Dits et écrits, I*, p. 235; *Essential Works of Foucault*, 2:71.

38. In chap. 9 of *Words and Things*, "Man and His Doubles," Foucault says that man is within a power that disperses him, the power of his own being. See MC 345 / 334–35.

39. On the connection of madness to animality, see Foucault, *Histoire de la folie à l'âge classique*, pp. 198–203.

40. See Foucault, *"Society Must Be Defended,"* p. 24 / 27. Here Foucault speaks of the distinction between the king and sovereignty in a single edifice and multiple subjugations within the social body.

41. NC x / xiv. Already in *The Birth of the Clinic*, in 1963, the time of "Preface to Transgression," Foucault describes the eye and power (*pouvoir*) as productive. See also Foucault, *Discipline and Punish*, pp. 30–32 / 23–24.

9. "This Is What We Must Not Do"

1. See Avishai Margalit, "The Suicide Bombers," *The New York Review of Books* 50, no. 1 (January 16, 2003). See also Dennis Keenan, *The Question of Sacrifice* (Albany: State University of New York Press, 2005).

2. An investigation would have to determine whether this thinking is part of bio-power or of sovereign power, or whether it constitutes a new form of power or reverts back to a very ancient one. See Michel Foucault, *Histoire de la sexualité, I: La Volonté de savoir* (Paris: Gallimard, 1976), esp. pp. 175–211; English translation by Robert Hurley as *The History of Sexuality: Volume I: An Introduction* (New York: Vintage, 1990), pp. 135–59.

3. Martin Heidegger, "Nietzsches Wort 'Gott ist tod,'" in *Holzwege* (Frankfurt am Main: Klostermann, 2003), pp. 209–67; English translation by William Lovitt as "The Word of Nietzsche: 'God Is Dead,'" in *The Question concerning Technology and Other Essays* (New York: Harper, 1977), pp. 53–112.

4. Michel Foucault, "Vie: Expérience et science," in *Dits et écrits IV, 1980–1988* (Paris: Gallimard, 1994), p. 776; English translation by Robert Hurley as "Life: Experience and Science," in *Essential Works of Michel Foucault, vol. 2, Aesthetics, Method, and Epistemology*, ed. James D. Faubion (New York: The New Press, 1998), p. 477. The original publication date of "Life" is 1984. See also Thierry Hoquet's excellent introductory essay to *La Vie* (Paris: Flammarion, 1999), pp. 11–41, and *Notions de philosophie, I*, "Le Vivant" (Paris: Gallimard, 1995), pp. 231–300.

5. Maurice Merleau-Ponty, *La Nature, notes cours du Collège de France*, ed. Dominique Séglard (Paris: Seuil, 1995); English translation by Robert Vallier as *Nature: Course Notes from the Collège de France* (Evanston, Ill.: Northwestern University Press, 2003); hereafter cited with the abbreviation N, with reference first to the French, then to the English translation.

6. The term *archeology* does not appear in the nature lectures until the third course, in the academic year 1959–60 (N 268, 273). Yet it is the first course, from the academic year 1956–57, that most resembles an archeology. For archeology in Merleau-Ponty, see: N 335 / 268, N 340 / 273; NdC 389; S 208 / 165. See also A. Hesnard, *L'Œuvre de Freud*, preface by Maurice Merleau-Ponty (Paris: Payot, 1960), p. 9; collected in Maurice Merleau-Ponty, *Parcours deux, 1951–1961* (Lagrasse: Verdier, 2000), p. 282; Maurice Merleau-Ponty, *Husserl at the Limits of Phenomenology*, trans. Leonard Lawlor with Bettina Bergo (Evanston, Ill.: Northwestern University Press, 2001), pp. BN 29note, BN 44; hereafter cited with the abbreviation HL, with reference to the BN numbers. See also my "The Chiasm and the Fold: An Introduction to the Philosophical Concept of Archeology," in *Thinking through French Philosophy* (Bloomington: Indiana University Press, 2003).

7. See Étienne Bimbinet, *Nature et humanité: Le Problème anthropologique dans l'œuvre de Merleau-Ponty* (Paris: Vrin, 2004), p. 228. See also Renaud Barbaras, "Merleau-Ponty et la nature," *Chiasmi International* 2 (2000): 58, speaking of a "dynamic morphology."

8. See Ronald Bruzina's "Method and Materiality in the Phenomenology of Subjectivity," *Philosophy Today*, SPEP Supplement, 43 (1997): 127–33, esp. p. 131.

9. Even in the 1958–59 lecture course "Contemporary Philosophy," where he lectures extensively on Heidegger, Merleau-Ponty never mentions death. Françoise Dastur, in her excellent essay on Merleau-Ponty and Heidegger, does not recognize this fact. See "Lecture de Heidegger," in Françoise Dastur, *Chair et langage: Essais sur Merleau-Ponty* (La Versanne: Encre Marine, 2001), pp. 191–217. See also, however, Françoise Dastur, *La mort* (Paris: Hatier, 1994), where she stresses, on p. 75, Merleau-Ponty's phrase "operative finitude" from *The Visible and the Invisible* (see VI 305 / 251).

10. See *Parcours Deux*, p. 371.

11. Deleuze begins his examination of Spinoza by referring to this passage from Merleau-Ponty. See Gilles Deleuze, *Spinoza et le problème de l'expression* (Paris: Minuit, 1968), p. 22; English translation by Martin Joughin as *Expressionism in Philosophy: Spinoza* (New York: Zone Books, 1990), p. 28.

12. In *Words and Things*, Foucault describes the exact relation to the infinite that Merleau-Ponty here is describing. Foucault says that the relation to the infinite in the classical epoch (Cartesianism), was a "negative relation." See MC 327 / 316.

13. See Chapter 6 of this book and its earlier version as "Man and His Doubles: Merleau-Ponty's Mixturism," in *Between Description and Interpreta-*

tion: The Hermeneutic Turn in Phenomenology, ed. Andre Wiercinski (Toronto: The Hermeneutic Press, 2005), pp. 125–38.

14. To conceive nature as produced by an artisan is, according to Merleau-Ponty, to engage in a kind of anthropologism (N 27 / 10).

15. Merleau-Ponty had already asserted this connection in 1946 in "Le Primat de la perception et ses conséquences philosophiques," *Bulletin de la société française de philosophie* 41 (December 1947): 135 and 151; English translation by James Edie as "The Primacy of Perception and Its Philosophical Consequences," in Maurice Merleau-Ponty, *The Primacy of Perception* (Evanston, Ill.: Northwestern University Press, 1964), pp. 27 and 41.

16. Deleuze, *Expressionism in Philosophy*, pp. 153 / 169.

17. Here Merleau-Ponty follows the phenomenological approach he laid out in *Phenomenology of Perception*, where essences must be grounded in existence. See Maurice Merleau-Ponty, *Phénoménologie de la perception* (Paris: Gallimard, 1945), pp. viii–ix; English translation by Colin Smith as *Phenomenology of Perception* (London: Routledge & Kegan Paul, 1962, revised 1979), pp. xiv–xv.

18. To use a phrase that Françoise Dastur has brought to our attention, Merleau-Ponty's thought is a thought of the inside, that is, a thought of immanence (Françoise Dastur, "La Pensée du dedans," in *Chair et language: Essais sur Merleau-Ponty* [La Versanne: Encre Marine, 2001], pp. 125–38). But we must not forget that Merleau-Ponty rejects Spinozistic monism as much as Cartesian dualism. Of course, Spinozistic monism is one of the models for Deleuze's idea of immanence.

19. See N 201 / 151; cf. VI 153 / 114; N 290 / 227.

20. See also Merleau-Ponty's description of Cézanne: "[Cézanne] thought himself powerless because he was not omnipotent, because he was not God, and wanted nevertheless to paint the world, to change it completely into a spectacle, to make *visible* how the world *touches* us" (Merleau-Ponty's italics). See Maurice Merleau-Ponty, "Le Doute de Cézanne," in *Sens et non-sens* (Paris: Gallimard, 1996 [1966]), p. 25; English translation by Hubert L. Dreyfus and Patricia Allen Dreyfus as "Cezanne's Doubt," in *Sense and Non-Sense* (Evanston, Ill.: Northwestern University Press, 1964), p. 19.

21. See N 61 / 38, S 225 / 178. For an excellent comparison of Merleau-Ponty's later thought with that of Schelling, see Robert Vallier, "*Être sauvage* and the Barbaric Principle: Merleau-Ponty's Reading of Schelling," *Chiasmi International* 2 (2000): 83–106.

22. This phrase appears already in Merleau-Ponty, "The Primacy of Perception and Its Philosophical Consequences," p. 134 / 26.

23. See Gilles Deleuze, *Différence et répétition* (Paris: Presses Universitaires de France, 1968), p. 90; English translation by Paul Patton as *Difference and Repetition* (New York: Columbia University Press, 1994), p. 64.

24. See N 64–65 / 40–41, NdC 1959–61 199, VI 30 / 14, VI 179 / 136, VI 281 / 228.

25. See N 279 / 217, VI 199 / 152; N 208 / 156, VI 319 / 265. On the idea of possibility in Merleau-Ponty, see David Morris, "What Is Living and What Is Non-Living in Merleau-Ponty's Philosophy of Movement and Expression," forthcoming in *Chiasmi International*, vol. 7.

26. For more on the "continuist presupposition," see Jacques Derrida, *Le toucher—Jean-Luc Nancy*.

27. Here we should recall the final pages of "Eye and Mind," where Merleau-Ponty stresses that there is no thought without a support.

28. See Gary Brent Madison, *The Phenomenology of Merleau-Ponty* (Athens: Ohio University Press, 1981), pp. 212–16.

29. Already in *The Structure of Behavior*, Merleau-Ponty speaks of the inertia of the living. See SC 239 / 222.

30. Again in *The Structure of Behavior*, Merleau-Ponty struggles against the retrospective illusion. See SC 225 / 218.

31. See John Sallis's excellent article on time in the *Phenomenology of Perception*: "Time, Subjectivity, and the Phenomenology of Perception," *The Modern Schoolman* 43 (May 1971): 343–57, esp. p. 352.

32. Maurice Merleau-Ponty, *L'Institution, la passivité, Notes de cours au Collège de France (1954–1955)* (Paris: Belin, 2003), p. 256; hereafter cited as IP.

33. As always, Paul Ricœur's interpretation of Freud is immensely helpful in determining Merleau-Ponty's understanding of Freud. See Paul Ricœur, *Freud and Philosophy*, trans. Denis Savage (New Haven: Yale University Press, 1970), bk. 2, chaps. 2 and 3, all of bk. 3. On p. 417, Ricœur refers his interpretation back to Merleau-Ponty.

34. See also Maurice Merleau-Ponty, *Éloge de la philosophie* (Paris: Gallimard, 1960), pp. 75–76; English translation by John Wild, James Edie, and John O'Neill as *In Praise of Philosophy and Other Essays* (Evanston, Ill.: Northwestern University Press, 1988), pp. 64–65. Here, commenting on Lavalle, Merleau-Ponty says that death is incorporated into our souls. But the "incorporation" means essence becomes existence or "active becoming." "Active becoming" is then understood as the "eternal [or the absolute] becoming fluid and flowing back from the end into the heart of life." Merleau-Ponty completes this note by saying that "The same fragile principle makes us alive and also gives to what we do a sense that does not wear out." In other words, death is not deprived of sense. See also the 1951 "Man and Adversity" (in *Signs*), where Merleau-Ponty makes awakening from the dispersion of sleep as the awakening of a meaning from beyond the grave the contrary of the undoing of meaning in the throes of the death agony (S 230–31 / 292).

35. See again Merleau-Ponty, "The Primacy of Perception and Its Philosophical Consequences," p. 123 / 15.

36. NC 148 / 145. The English translation is based on the 1972 edition, which contains many changes from the original 1963 edition, but it also at times inserts passages from the 1963 edition. As a result, the English trans-

lation is a confusing text. Concerning the revisions that Foucault made for the 1972 edition, see David Macey, *The Lives of Michel Foucault* (New York: Vintage, 1995), p. 133.

37. With the idea of non-sense (not absurdity), we would be able to introduce the ideas of genuine singularity and genuine multiplicity.

38. For this conception of the animal in terms of traces, see Jacques Derrida, "L'Animal que donc je suis (à suivre)," in *L'Animal autobiographique: Autour de Jacques Derrida* (Paris: Galilée, 1999), pp. 251–302; English translation by David Wills as "The Animal That Therefore I Am (More to Follow)," *Critical Inquiry* 28 (Winter 2002): 370–418. Derrida's lectures on animality were originally presented in 1998.

39. One finds both of these images, carrying (*porter*) and following (*suivre*), in Derrida.

40. Maurice Merleau-Ponty, *Les Aventures de la dialectique* (Paris: Gallimard, 1955), p. 228; English translation by Joseph Bien as *Adventures of the Dialectic* (Evanston, Ill.: Northwestern University Press, 1973), p. 164.

41. Michel Foucault, *Surveiller et punir* (Paris: Gallimard, 1975), p. 197; English translation by Alan Sheridan as *Discipline and Punish* (New York: Vintage, 1977), p. 168.

10. Metaphysics and Powerlessness

1. Michel Foucault, *Histoire de la sexualité, I: La Volonté de savoir* (Paris: Gallimard, 1976), esp. pp. 175–211; English translation by Robert Hurley as *The History of Sexuality: Volume I: An Introduction* (New York: Vintage, 1990), pp. 135–59. Hereafter cited as HS1, with reference first to the French, then to the English translation.

2. For an interesting discussion of this case, see John Protevi, "The Schiavo Case: Jurisprudence, Biopower, and Privacy as Singularity," unpublished manuscript.

3. See my "*Verendlichung* (Finitization): The Overcoming of Metaphysics with Life," in *Philosophy Today* (Winter 2004): 390–403.

4. As is well known, Heidegger argues that Nietzsche's will to power is the "completion" of Western metaphysics understood as the history of nihilism (Martin Heidegger, "Nietzsches Wort 'Gott ist tod,'" in *Holzwege* [Frankfurt am Main: Klostermann, 2003], p. 259; English translation by William Lovitt as "The Word of Nietzsche: 'God Is Dead,'" in *The Question concerning Technology and Other Essays* [New York: Harper, 1977], p. 104). This text will be abbreviated hereafter as NW, with the German page number preceding the English page number. In *The History of Sexuality, Volume I*, Foucault argues that bio-power is a "mutation" of juridical power, the kind of power that functioned in the classical epoch (HS1 155 / 117). These two words, *completion* (*Vollendung*) and *mutation*, indicate a difference in how Heidegger and Foucault conceive history.

5. The interpretation that I will present is based mainly on Heidegger's 1943 essay, "Nietzsches Wort 'Gott ist tod.'" I am privileging this essay be-

cause Heidegger tells us that it is based on his final lecture courses on Nietzsche from 1936 to 1940; see *Holzwege*, pp. 375–76; *The Question concerning Technology*, p. x. In fact, the essay seems closely related to two lecture courses, "The Will to Power as Knowledge," delivered in 1939, and "European Nihilism," delivered in 1940. See Martin Heidegger, "Der Wille zur Macht als Erkenntnis," in *Nietzsche I* (Pfullingen: Neske, 1961), pp. 475–658; English translation by Joan Stambaugh, David Farrell Krell, and Frank A. Capuzzi as "Part One: The Will to Power as Knowledge," in *Nietzsche*, vol. 3, *The Will to Power as Knowledge and as Metaphysics* (New York: Harper and Row, 1987), pp. 1–158. See also Martin Heidegger, "Der Europäische Nihilismus," in *Nietzsche II* (Pfullingen: Neske, 1961), pp. 31–256; English translation by Frank A. Capuzzi as "Part One: European Nihilism," in *Nietzsche*, vol. 4, *Nihilism* (New York: Harper and Row, 1982), pp. 1–196. For the discussion of valuation as preservation-enhancement conditions, see *Nietzsche* 2:96–109 / 4:58–68. For the discussion of command, see *Nietzsche* 1:606–16 / 3:115–122. For reversing Platonism, see *Nietzsche* 2:199–202 / 4:147–49. For modern subjectivism, see *Nietzsche* 2:141–68 / 4:96–118. While Heidegger repeatedly returns to the biologistic reading of Nietzsche in the lecture, the first and perhaps decisive discussion occurs in "Der Wille zur Macht als Kunst," in *Nietzsche I*; English translation by David Farrell Krell as *Nietzsche*, vol. 1, *The Will to Power as Art* (New York: Harper and Row, 1979). There Heidegger says that " 'Life' here means neither mere animal and vegetable being nor that readily comprehensible and compulsive busyness of the everyday; rather, 'life' is the term for being in its new interpretation according to which it is a becoming. 'Life' is neither 'biologically' nor 'practically' intended; it is meant metaphysically. The equation of being and life is not some sort of unjustified expansion of the biological, although it often seems that way, but a transformed interpretation of the biological on the basis of being, grasped in a superior way—this, of course, not fully mastered, in the timeworn schema of 'being and becoming' " (*Nietzsche* 1:253 / 1:219).

6. The interpretation that I am presenting is based mainly on Michel Foucault, *The History of Sexuality: Volume 1: An Introduction* and Michel Foucault, *Surveiller et punir* (Paris: Gallimard, 1975); English translation by Alan Sheridan as *Discipline and Punish* (New York: Vintage, 1995). Hereafter cited as SP with reference first to the original French, then to the English translation. Foucault delivered lecture courses at the Collège de France during this period, which shed a lot of light on the genealogy and mechanisms of modern bio-power. See Michel Foucault, *Les Anormaux, Cours au Collège de France, 1974–1975* (Paris: Gallimard and Seuil, 1999); English translation by Graham Burchell as *Abnormal, Lectures at the Collège de France, 1974–1975* (New York: Picador, 2004); also, Michel Foucault, *"Il faut défendre la société," Cours au Collège de France, 1976* (Paris: Gallimard and Seuil, 1997); English translation by David Macey as *"Society Must Be Defended," Lectures at the Col-*

lège de France, 1975–76 (New York: Picador, 2003). For a brief assessment of the relation of Foucault to Heidegger, see Dominique Janicaud, *Heidegger en France, I: Récit* (Paris: Albin Michel, 2001), pp. 215–16.

7. Martin Heidegger, *Gesamtausgabe*, vol. 9: *Wegmarken* (Frankfurt am Main: Klostermann, 1976); English translation edited by William McNeill as *Pathmarks* (Cambridge: Cambridge University Press, 1998). Hereafter cited as GA9, with reference to the 1976 edition, then the English translation.

8. See also Michel Foucault, "Nietzsche, la généalogie, l'histoire," in *Dits et écrits I, 1954–1975* (Paris: Gallimard, 2001), p. 1015; English translation as "Nietzsche, Genealogy, History" by Donald F. Bourchard and Sherry Simon in *Essential Works of Foucault, 1954–1984*, vol. 2: *Aesthetics, Method, Epistemology* (New York: The New Press, 1998), p. 380. For Heidegger, following Nietzsche, anti-Platonism is the movement of the devaluation of the highest values (nihilism): God, the Good, the Beautiful, the supersensory in general, the moral law, all of these values and ideals are devaluing themselves (NW 217/61). In his 1971 "Nietzsche, Genealogy, History," Foucault opposes what he calls "actual history" to Platonism, saying that actual history "hollows out that upon which we like to make history rest," that is, the Ideas, the moral law, spirit, and, importantly, the immortality of the soul. While the word *Platonism* does not appear in *The History of Sexuality, Volume I*, the word *Christianity* appears repeatedly. There Foucault says that the new techniques of sexuality are continuous with certain Christian practices found in the Middle Ages. Nevertheless, this continuity "did not prevent a major transformation" (HS1 155/117). The technology of sex was no longer concerned with "death and everlasting punishment"; instead, at "the end of the Eighteenth Century," "the flesh is brought down to the level of [*est rabbatue sur*] the organism" (HS1 155/117). See also Foucault, *"Society Must Be Defended," Lectures at the Collège de France, 1975–1976*, pp. 153–54/172–73, where Foucault opposes historicism to the "Platonic" idea that knowledge must belong to the register of order and peace.

9. The idea of the Christ starts the movement of anti-Platonism.

10. The transformation, in the West, is a transformation in Christianity. In fact, the resistance to bio-will to power has to consist in what Jean-Luc Nancy has called "the deconstruction of Christianity." See Jean-Luc Nancy, "La Déconstruction du christianisme," *Les Études Philosophiques*, no. 4 (1998): 503–19; translated by Simon Sparks as "The Deconstruction of Christianity," in *Religion and Media*, ed. Hent de Vries and Samuel Weber (Stanford: Stanford University Press, 2001), pp. 112–30.

11. "Appraisal" translates *Schätzung*. When Heidegger says here *"another appraisal of life,"* he is referring to the Platonistic appraisal of life. In Platonism (as in Christianity), the supersensory world or God determines earthly life from above, beyond, or from the outside of life (NW 220/64).

12. The idea that life is valuation can also be found in Georges Canguilhem, *Le Normal et le pathologique* (Paris: Presses Universitaires de France,

2003 [1966]), p. 134; English translation by Carolyn R. Fawcett in collaboration with Robert S. Cohen as *The Normal and the Pathological* (New York: Zone Books, 1991), p. 201. Nietzsche, as Foucault points out in "Life: Experience and Science," influences Canguilhem here. See also Guillaume Leblanc, *Canguilhem et les norms* (Paris: Presses Universitaires de France, 1998).

13. For the connection between the two texts, see HS1 185 / 140.

14. The Panopticon would be shaped in a ring; the prisoners would be separated into cells that would be arranged across the peripheral circle; each cell would have a window in the outside wall so that light could enter the cell, making the prisoner visible, and a window in the inside wall so that an invisible guard located in the central tower could oversee all the prisoners (*surveiller*). The Panopticon, for Foucault, unifies the techniques by means of which earlier ages had dealt with lepers (exclusion) and with the plague (multiple separations) (SP 231 / 198).

15. With regard to this claim, we should not overlook Foucault's constant interest in the theoretical or speculative role that the eighteenth-century Ideologues (Condillac, in particular) play in the development of the modern clinic (NC 96 / 96), of modern humanism, and of the modern regime of power (HS1 184 / 140). See MC 77 / 63.

16. *Bestand* is the word, of course, that Heidegger uses to designate the mode of presencing in modern technology; see Martin Heidegger, *Die Technik und die Kehre* (Stuttgart: J. G. Cotta'sche, 1962), p. 16; "The Question concerning Technology," in *The Question Concerning Technology and Other Essays*, p. 17.

17. The will to power is not the ordering about of others, and it is, according to Heidegger, "more difficult than obeying" (NW 234 / 77).

18. "Exacting" translates *prélèvement*.

19. It is quite important that Foucault's purpose in "Nietzsche, Genealogy, History" consists in showing how Nietzsche's later thought differs from his earlier thought, especially in regard to "the power [*pouvoir*] of life to affirm and create." See *Dits et écrits I, 1954–1975*, p. 1024; *Essential Works of Foucault, 1954–1984*, 2:388.

20. Foucault, of course, had been interested in the transformation that Descartes represents as early as *L'Histoire de la folie à l'âge classique*, in 1961. See Michel Foucault, *L'Histoire de la folie à l'âge classique* (Paris: Gallimard, 1972), esp. pp. 67 and 437. Foucault revised the 1961 book (like *Naissance de la clinique*) for the 1972 edition. There is an English translation of an abridged French edition called *Madness and Civilization*, trans., Richard Howard (New York: Vintage, 1973).

21. Heidegger says, "The un-positing [*Absetzung*] of the supersensory ends in a 'neither-nor' in relation to the *difference* between the sensory (*aestheton*) and the non-sensory (*noeton*). The un-positing ends in meaninglessness [*Sinnlosen*]. Nevertheless, the un-positing of the supersensory remains

the unthought and invincible presupposition of the blind attempts to extricate themselves from meaninglessness [*Sinnlosen*] through mere sense-donation [*Sinn-gebung*]" (NW 209 / 54, my emphasis). "Sense-donation" is how Husserl defines intentionality in *Ideas I*, paragraph 55. With regard to the connection between Husserlian phenomenology and Nietzsche's thought of the will to power, it is important to keep in mind the centrality of the concept of *Geltung* ("validity") and *Wert* ("value") in Husserl's thought. See Françoise Dastur, *La Phénoménologie en questions* (Paris: Vrin, 2004), p. 29.

22. For the ambiguity between *noesis* and *noema*, see Edmund Husserl, Hua IX: *Phänomenologische Psychologie* (The Hague: Martinus Nijhoff, 1962), p. 292; English translation by Richard E. Palmer in *The Essential Husserl*, ed. Donn Welton (Bloomington: Indiana University Press, 1999), p. 331. See also Eugen Fink, "Die Phänomenologische Philosophie E. Husserl in der Gegenwärtigen Kritik," originally published in *Kantstudien* 38, no. 3/4 (Berlin, 1933); collected in Eugen Fink, *Studien zur Phänomenologie* (The Hague: Martinus Nijhoff, 1966), pp. 132–33; English translation as "The Phenomenological Philosophy of Edmund Husserl and Contemporary Criticism," *The Phenomenology of Husserl*, ed. R. O. Elveton (Chicago: Quadrangle Books, 1970), pp. 124–25. For *reell* versus *irreel*, see Edmund Husserl, Hua III.1: *Ideen zu einer reinen Phänomenologie und phänomenologischen Philosophie*, bk. 1, ed. Karl Schuhmann (The Hague: Martinus Nijhoff, 1976), pp. 55–57 (paragraphs 31–32); English translation by F. Kersten as *Ideas pertaining to a Pure Phenomenology and to a Phenomenological Philosophy* (The Hague: Martinus Nijhoff, 1982), pp. 59–61. See also Edmund Husserl, *Idées directrices pour une phénoménologie*, trans. Paul Ricœur (Paris: Gallimard, 1950). For Husserl's use of the word *reality*, see paragraph 42.

23. Martin Heidegger, *Kant und das Problem der Metaphysik*, 4th, expanded ed. (Frankfurt am Main: Klostermann, 1973), p. 182 (paragraph 34); English translation by Richard Taft as *Kant and the Problem of Metaphysics*, 4th ed., enlarged (Bloomington: Indiana University Press, 1990), p. 129.

24. We must note the homonymic relation, in French, of *et* and *est* ("is"): *sum moribundus*.

25. In relation to the *écart*, it is necessary to re-read the paradoxes that phenomenology entails, according to Fink. The paradoxical nature of the *écart* (as well as blindness) is why one can speak of life as a question or a problem (or problematization). See Fink, "The Phenomenological Philosophy of Edmund Husserl and Contemporary Criticism," pp. 151–56 / 140–45.

26. In many regards, the critique of phenomenology and Heidegger that one finds in Foucault resembles the one that one finds in Levinas. See, in particular, "Is Ontology Fundamental," English translation in *Emmanuel Levinas: Basic Philosophical Writings*, ed. Adriaan T. Peperzak, Simon Critchley, and Robert Bernasconi (Bloomington: Indiana University Press, 1996), esp. pp. 3–5, where Levinas speaks of the "ambiguity of contempo-

rary ontology" and seems to be attempting to differentiate beings from being, from being mixed with being. Levinas too, of course, focuses on the "haecceity," the "this-ness" of beings, in other words, their singularity, or their alterity. He too thinks that they cannot ultimately be subsumed under general concepts or representations, or understood. Also, when Levinas criticizes Heidegger later, he focuses on the idea of gathering (*Versammlung*). See Emmanuel Levinas, *Autrement qu'être ou au-delà de l'essence* (The Hague: Martinus Nijhoff, 1974), p. 220; English translation by Alphonso Lingis as *Otherwise than Being or Beyond Essence* (The Hague: Martinus Nijhoff, 1981), p. 140. Finally, see also Appendix One to my *Thinking through French Philosophy* (Bloomington: Indiana University Press, 2003), p. 152.

27. Derrida calls this difference "différence." It is here that we have the entire problem of repetition and memory.

28. For more on spacing, see Jacques Derrida, *Le toucher — Jean-Luc Nancy* (Paris: Galilée, 2000).

29. See Martin Heidegger, *Gesamtausgabe*, vol. 29/30, *Die Grundbegriffe der Metaphysik: Welt — Endlichkeit — Einsamkeit* (Frankfurt am Main: Klostermann, 1983), p. 396; translated by William McNeill and Nicholas Walker as *The Fundamental Concepts of Metaphysics: World, Finitude, Solitude* (Bloomington: Indiana University Press, 1995), p. 273. Hereafter cited as GA29/30, with reference first to the German, then to the English. For more on "What Is Metaphysics?" especially the audience's reaction to Heidegger's address, see also Rudiger Safranski, *Martin Heidegger: Between Good and Evil*, trans. Ewald Osers (Cambridge: Harvard University Press, 1998). Moreover, we would have to distinguish this *Lebensprozesses* from what Heidegger calls "der Grundvorgang des Lebens," that is, valuation (*Wertung*) and the securing of permanence. See Heidegger, *Nietzsche*, 1:544 and 591 / 3:61 and 101. For a comprehensive summary of Heidegger's relation to biologism, see Robert Bernasconi's "Heidegger, Nietzsche, and the Critique of Biologism" (unpublished manuscript).

30. On indifference in the early Heidegger, see Theodore Kiesiel, *The Genesis of Heidegger's Being and Time* (Berkeley: University of California Press, 1993), p. 36.

31. I return to the discussion of death in *Being and Time* below, at the end of this section.

32. See Jean-Luc Marion, *Réduction et donation* (Paris: Presses Universitaires de France, 1989), pp. 11–18; English translation by Thomas A. Carlson as *Reduction and Givenness* (Evanston, Ill.: Northwestern University Press, 1998), pp. 71–76. Here Marion proposes a short reading of "What Is Metaphysics?" as a "reduction to the Nothing." While Marion stresses the indifference of the experience of anxiety, he does not mention anxiety as the experience of powerlessness.

33. Cf. Martin Heidegger, "Die Sprache," in *Unterwegs zur Sprache* (Pfullingen: Neske, 1982 [1959]), p. 24; English translation by Albert Hofstadter

as "Language," in *Poetry, Language, Thought* (New York: Harper & Row, 2001), p. 199. Here, describing the intimacy of world and thing, Heidegger says: "The intimacy of world and thing is no mixture [*keine Verschmelzung*]. Intimacy prevails [*waltet*] only where the intimate, world and thing, purely divides itself and remains different. [*Innigkeit walte nur, wo das Innige, Welt und Ding, rein sich scheidet und geschieden bleibt.*]"

34. See also Heidegger's use of this term in "On the Essence of Truth" (GA9 184–85 / 141–42). Cf. also Martin Heidegger, *Contributions to Philosophy*, trans. Parvis Emad and Kenneth Maly (Bloomington: Indiana University Press, 1999), p. 6: "This thinking-saying is a directive [*eine Weisung*]. It indicates the free sheltering of the truth of be-ing in beings as a necessity, without being a command." It seems that *Weisung* refers to the idea of a horizon; see Heidegger, *Nietzsche*, 1:589 / 3:99.

35. See Jacques Derrida, "L'Oreille de Heidegger," in *Politiques de l'amitié* (Paris: Galilée, 1994), p. 349; English translation by John P. Leavey, Jr., as "Heidegger's Ear," in *Reading Heidegger*, ed. John Sallis (Bloomington: Indiana University Press, 1993), p. 168.

36. Derrida says that "The relation to the other is deference itself." See Jacques Derrida, *Adieu à Emmanuel Levinas* (Paris: Galilée, 1997), p. 88; English translation by Pascale-Anne Brault and Michael Naas as *Adieu to Emmanuel Levinas* (Stanford: Stanford University Press, 1999), p. 46.

37. On indifference in the early Heidegger, see Theodore Kiesiel, *The Genesis of Heidegger's Being and Time* (Berkeley: University of California Press, 1993), p. 36.

38. Martin Heidegger, *Gesamtausgabe*, vol. 26: *Metaphysische Anfangsgründe der Logik im Ausgang von Leibniz* (Frankfurt am Main: Klosterman, 1978), p. 272; English translation by Michael Heim as *The Metaphysical Foundations of Logic* (Bloomington: Indiana University Press, 1984), p. 210.

39. See Heidegger, *Nietzsche*, 1:541 / 3:58.

40. Gadamer, *Heidegger's Ways*, pp. 46–47.

41. It is possible to claim that Heidegger conceives the ontological difference as the same through the concept of gathering (*Versammlung*). On *Versammlung*, see Jacques Derrida, "La Main de Heidegger (*Geschlecht* II) (1984–1985)," in *Psyché* (Paris: Galilée, 1987), pp. 415–52; English translation by John P. Leavey, Jr., as "Heidegger's *Geschlecht* II: Heidegger's Hand," in *Deconstruction and Philosophy*, ed. John Sallis (Chicago: University of Chicago Press, 1987), pp. 161–98.

42. McNeill and Walker render *Verendlichung* as "becoming finite" (without the hyphen). See GA29/30 8 / 6. See also William McNeill, *The Glance of the Eye* (Albany: State University of New York Press, 1999), p. 119, where McNeill connects *Verendlichung* with individuation and the *Augenblick*. See also Françoise Dastur, "Mortalité et finitude," in *La Phénoménologie en questions* (Paris: Vrin, 2004), p. 237. One could say that all of the essays that Dastur has collected in this volume concern finitization.

43. See also Jean Greisch, *Ontologie et temporalité: Esquisse d'une interpréta-tion intégrale de Sein und Zeit* (Paris: Presses Universitaires de France, 1994), p. 273n1. This finitization of freedom seems to anticipate the concept of freedom in "On the Essence of Truth": man does not possess freedom; free-dom possesses man (see GA9 85 / 145). It also seems to be connected to the following comment from "On the Essence of Grounds": "The fact that the ever-excessive projection of world attains its power and becomes our pos-session only in such withdrawal is at the same time a transcendental testi-mony to the *finitude* [Endlichkeit] of Dasein's freedom. And does not the *finite* essence of freedom in general thereby announce itself?" (GA9 63 / 129, Heidegger's emphasis).

44. Martin Heidegger, *Sein und Zeit* (Tübingen: Max Niemeyer, 1979), p. 245, section 48; English translation by Joan Stambaugh as *Being and Time* (Albany: State University of New York Press, 1996), p. 228.

45. Ibid., 247 / 229.

46. Ibid., 261 / 241.

47. See also Jacques Derrida, *Apories* (Paris: Galilée, 1996), pp. 60–63; English translation by Thomas Dutoit as *Aporias* (Stanford: Stanford Uni-versity Press, 1993), pp. 30–31. The English translation was made from an earlier version, found in *Le Passage des frontières* (Paris: Galilée, 1993).

48. In this regard, death is something positive, so to speak, a "negative life." See Michel Foucault, "Préface à la transgression," in *Dits et ecrits I, 1954–1975* (Paris: Gallimard, 2001), pp. 261–78, esp. p. 266; English transla-tion by Donald F. Bourchard as "Preface to Transgression," in *Essential Works of Foucault, vol. 2: Aesthetics, Method, and Epistemology*, ed. James D. Faubion (New York: The New Press, 1998), pp. 69–87, esp. p. 74. On this page, Foucault cites Kant's early essay on negative magnitudes; on the basis of how negative numbers function in relation to positive numbers, Kant stresses that real conflict occurs only between two positive forces. See Im-manuel Kant, *Theoretical Philosophy 1755–1770*, ed. David Walford (Cam-bridge: Cambridge University Press, 1992), pp. 206–41, esp. p. 215.

49. Gilles Deleuze, *Foucault* (Paris: Minuit, 1986), p. 129; English trans-lation by Seán Hand as *Foucault* (Minneapolis: University of Minnesota Press, 1988), p. 121. According to Deleuze, "Bichat's zone" is "an outside, an atmospheric element, a 'non-stratified substance' that would be capable of explaining how the two forms of knowledge can embrace and intertwine on each stratum, from one edge of the fissure to the other. . . . This informal outside is a battle, a turbulent, stormy zone where singular points and the relation of forces between these points are tossed about. Strata merely col-lected and solidified the visual dust and the sonic echo of the battle raging above them. But, up above, the singularities have no form and are neither visible bodies nor speaking persons. We enter into the domain of uncertain doubles and partial deaths, a domain of emergence and vanishing (Bichat's zone). This is a micro-physics" (translation modified). According to De-

leuze, for Foucault, "Bichat broke with the classical conception of death, as a decisive moment or indivisible event, and broke with it in two ways, simultaneously presenting death as being co-extensive with life and as something made up of singular deaths. When Foucault analyses Bichat's theories, his tone demonstrates sufficiently that he is concerned with something other than an epistemological analysis: he is concerned with a conception of death (102 / 95; translation modified). Deleuze makes one other significant mention of Bichat in relation to Foucault: "it is Bichat who breaks with the classical conception of death, as being a decisive, indivisible instant. . . . Bichat's three great innovations are to have posited death as being co-extensive with life, to have made it the global result of partial deaths, and above all to have taken 'violent death' rather than 'natural death' as the model. . . . Bichat's book [*Recherches physiologiques sur la vie et la mort*] is the first act of a modern conception of death" (138n12 / 152n12). In reference to this zone, we should also keep in mind that Merleau-Ponty, at the end of *L'Œil et l'esprit*, speaks of a "zone of the fundamental" (OE 91 / 149). Also on Foucault and Bichat, see John Rajchman, *Truth and Eros: Foucault, Lacan and the Question of Ethics* (New York: Routledge, 1991), p. 35.

50. Deleuze, *Foucault*, 98–99 / 93.

51. This entire discussion implies that Foucault, following Bichat, is entering into a zone that is different from what Heidegger, in *Being and Time*, calls present-at-hand or ready-to-hand. See *Being and Time*, p. 238 / 221. Here Heidegger speaks of the corpse and the student of pathological anatomy.

52. Derrida also stresses this connection of mortalism and vitalism in Foucault's study of Bichat. See Jacques Derrida, "'To Do Justice to Freud': The History of Madness in the Age of Psychoanalysis," trans. Pascale-Anne Brault and Michael Naas, in *Foucault and His Interlocutors*, ed. Arnold Davidson (Chicago: University of Chicago Press, 1990), pp. 85–86.

53. For more on this conception of death, see Deleuze, *Difference and Repetition*, pp. 148–49 / 112–13. See also Maurice Blanchot, "L'Œuvre et l'espace de la mort," in *L'Espace littéraire* (Paris: Gallimard, 1955), pp. 99–211; English translation by Ann Smock as "The Work and Death's Space," in *The Space of Literature* (Lincoln: University of Nebraska Press, 1982), pp. 101–59.

54. Canguilhem says that life contains a "superabundance" of solutions to problems that have not yet appeared, a superabundance that can be abused and lead to error. See *The Normal and the Pathological*, pp. 133, 199, and 206 / 200, 265, and 273.

55. See Deleuze, "Immanence: Une vie," in *Deux régimes de Fous: Textes et entretiens, 1975–1995* (Paris: Minuit, 2003), pp. 359–64; translation by Anne Boyman as *Pure Immanence: Essays on a Life* (New York: Zone, 2001), pp. 25–34. If we follow Deleuze, we would have to make a distinction in Foucault between a singularity and an ordinary point. The divergences between

the points amount to degeneration or *usure* or finitization; the points are ordinary. The disease, however, such as cancer, is an event among the ordinary points; this event is a remarkable point, a genuine singularity.

56. The title of Chapter 9 of *The Birth of the Clinic* is ambiguous: "L'invisible visible." The title could be translated either as "the visible invisible" or as "the invisible visible."

57. On "informal," see Michel Foucault, "La Pensée du dehors," in *Dits et écrits, I, 1954–1975* (Paris: Gallimard, 2001), p. 566; English translation by Brian Massumi as "The Thought from Outside," in *Foucault/Blanchot* (New York: Zone Books, 1997), p. 55. Deleuze calls informality "Bichat's zone." See Deleuze, *Foucault*, p. 129 / 121, and n. 51, above.

58. Foucault says that "fiction consists *not* in making us see the invisible, *but* in making us see *how much* the invisibility of the visible is invisible." See Foucault, "The Thought from Outside," p. 524 / 24, my emphasis.

59. Friedrich Nietzsche, *Götzen-Dämmerung*, in *Kritische Studienausgabe*, vol. 6 (Berlin: De Gruyter, 1988), pp. 80–81; English translation by Walter Kaufmann as *The Twilight of the Idols*, in *The Portable Nietzsche* (New York: Viking, 1968), p. 485.

60. For more on this transformation from a metaphysics of nature, see also Maurice Merleau-Ponty, "Le Langage indirect et les voix du silence," S 65 / 52. As we shall see, Foucault expresses this transformation most clearly. See NC 157 / 154. In 1972, Foucault revised this passage.

61. By means of the concept of aging, it is possible to claim that for Levinas too what is "otherwise than being" is life. See Levinas, *Otherwise than Being*, 86 / 51.

62. Heidegger would call this limit "the nothing." See Martin Heidegger, "Was ist Metaphysik?" in GA9, 113 / 90.

63. See Edmund Husserl, *L'Origine de la géométrie*, trans. and introd. Jacques Derrida (Paris: Presses Universitaires de France, 1974 [1962]), p. 108; English translation by John P. Leavey, Jr., as *Edmund Husserl's Origin of Geometry: An Introduction* (Lincoln: University of Nebraska Press, 1989 [1978]), p. 105. See also NC 201 / 197. For a more contemporary use of the phrase "finitude originaire," see Dastur, *La Phénoménologie en questions*, p. 104.

64. Cf. Jacques Lacan, "Subversion du sujet et dialectique du désir dans l'inconscient Freudian," in *Écrits* (Paris: Seuil, 1966), esp. p. 803; English translation by Alan Sheridan as "Subversion of the Subject and Dialectic of Desire," in *Écrits: A Selection* (New York: Norton, 1977), p. 301. Here Lacan speaks of Freud's biologism, whose true tone can be discovered only if you are "made to live the death instinct."

65. See also Gilles Deleuze, "A quoi reconnaît-on le structuralisme?" in *L'Île déserte et autres textes* (Paris: Minuit, 2002), p. 261; English translation by Melissa McMahon as "How Do We Recognize Structuralism?" in *Desert Islands and Other Texts* (New York: Semiotext(e), 2004), p. 186.

66. See Deleuze, *Foucault*, pp. 98–99 / 92–93. I think that *Foucault* is perhaps one of the greatest philosophy books written in the twentieth century. It is indisputable, in any event, that Deleuze has given us the most philosophically interesting reading of Foucault's diverse corpus.

Conclusion: The Followers

1. See my "Where Does Obscurity Come From? Mixturism in Paul Valéry's 'La Jeune Parque,'" in *Phenomenology and Literature*, ed. Pol Vandvelde (Würzberg: Koenigshausen und Neumann, forthcoming).

Index

Perspectives in Continental Philosophy Series

John D. Caputo, series editor

1. John D. Caputo, ed., *Deconstruction in a Nutshell: A Conversation with Jacques Derrida.*

2. Michael Strawser, *Both/And: Reading Kierkegaard—From Irony to Edification.*

3. Michael D. Barber, *Ethical Hermeneutics: Rationality in Enrique Dussel's Philosophy of Liberation.*

4. James H. Olthuis, ed., *Knowing* Other-*wise: Philosophy at the Threshold of Spirituality.*

5. James Swindal, *Reflection Revisited: Jürgen Habermas's Discursive Theory of Truth.*

6. Richard Kearney, *Poetics of Imagining: Modern and Postmodern.* Second edition.

7. Thomas W. Busch, *Circulating Being: From Embodiment to Incorporation—Essays on Late Existentialism.*

8. Edith Wyschogrod, *Emmanuel Levinas: The Problem of Ethical Metaphysics.* Second edition.

9. Francis J. Ambrosio, ed., *The Question of Christian Philosophy Today.*

10. Jeffrey Bloechl, ed., *The Face of the Other and the Trace of God: Essays on the Philosophy of Emmanuel Levinas.*

11. Ilse N. Bulhof and Laurens ten Kate, eds., *Flight of the Gods: Philosophical Perspectives on Negative Theology.*

12. Trish Glazebrook, *Heidegger's Philosophy of Science.*

13. Kevin Hart, *The Trespass of the Sign: Deconstruction, Theology, and Philosophy.*

14. Mark C. Taylor, *Journeys to Selfhood: Hegel and Kierkegaard*. Second edition.

15. Dominique Janicaud, Jean-François Courtine, Jean-Louis Chrétien, Michel Henry, Jean-Luc Marion, and Paul Ricœur, *Phenomenology and the "Theological Turn": The French Debate*.

16. Karl Jaspers, *The Question of German Guilt*. Introduction by Joseph W. Koterski, S.J.

17. Jean-Luc Marion, *The Idol and Distance: Five Studies*. Translated with an introduction by Thomas A. Carlson.

18. Jeffrey Dudiak, *The Intrigue of Ethics: A Reading of the Idea of Discourse in the Thought of Emmanuel Levinas*.

19. Robyn Horner, *Rethinking God as Gift: Marion, Derrida, and the Limits of Phenomenology*.

20. Mark Dooley, *The Politics of Exodus: Søren Keirkegaard's Ethics of Responsibility*.

21. Merold Westphal, *Toward a Postmodern Christian Faith: Overcoming Onto-Theology*.

22. Edith Wyschogrod, Jean-Joseph Goux and Eric Boynton, eds., *The Enigma of Gift and Sacrifice*.

23. Stanislas Breton, *The Word and the Cross*. Translated with an introduction by Jacquelyn Porter.

24. Jean-Luc Marion, *Prolegomena to Charity*. Translated by Stephen E. Lewis.

25. Peter H. Spader, *Scheler's Ethical Personalism: Its Logic, Development, and Promise*.

26. Jean-Louis Chrétien, *The Unforgettable and the Unhoped For*. Translated by Jeffrey Bloechl.

27. Don Cupitt, *Is Nothing Sacred? The Non-Realist Philosophy of Religion: Selected Essays*.

28. Jean-Luc Marion, *In Excess: Studies of Saturated Phenomena*. Translated by Robyn Horner and Vincent Berraud.

29. Phillip Goodchild, *Rethinking Philosophy of Religion: Approaches from Continental Philosophy*.

30. William J. Richardson, S.J., *Heidegger: Through Phenomenology to Thought*.

31. Jeffrey Andrew Barash, *Martin Heidegger and the Problem of Historical Meaning*.

32. Jean-Louis Chrétien, *Hand to Hand: Listening to the Work of Art*. Translated by Stephen E. Lewis.

33. Jean-Louis Chrétien, *The Call and the Response*. Translated with an introduction by Anne Davenport.

34. D. C. Schindler, *Han Urs von Balthasar and the Dramatic Structure of Truth: A Philosophical Investigation*.

35. Julian Wolfreys, ed., *Thinking Difference: Critics in Conversation*.

36. Allen Scult, *Being Jewish/Reading Heidegger: An Ontological Encounter.*

37. Richard Kearney, *Debates in Continental Philosophy: Conversations with Contemporary Thinkers.*

38. Jennifer Anna Gosetti-Ferencei, *Heidegger, Hölderlin, and the Subject of Poetic Language: Towards a New Poetics of Dasein.*

39. Jolita Pons, *Stealing a Gift: Kirkegaard's Pseudonyms and the Bible.*

40. Jean-Yves Lacoste, *Experience and the Absolute: Disputed Questions on the Humanity of Man.* Translated by Mark Raftery-Skehan.

41. Charles P. Bigger, *Between Chora and the Good: Metaphor's Metaphysical Neighborhood.*

42. Dominique Janicaud, *Phenomenology "Wide Open": After the French Debate.* Translated by Charles N. Cabral.

43. Ian Leask and Eoin Cassidy, eds. *Givenness and God: Questions of Jean-Luc Marion.*

44. Jacques Derrida, *Sovereignties in Question: The Poetics of Paul Celan.* Edited by Thomas Dutoit and Outi Pasanen.

45. William Desmond, *Is There a Sabbath for Thought? Between Religion and Philosophy.*

46. Bruce Ellis Benson and Norman Wirzba, eds. *The Phenomoenology of Prayer.*

47. S. Clark Buckner and Matthew Statler, eds. *Styles of Piety: Practicing Philosophy after the Death of God.*

48. Kevin Hart and Barbara Wall, eds. *The Experience of God: A Postmodern Response.*

49. John Panteleimon Manoussakis, *After God: Richard Kearney and the Religious Turn in Continental Philosophy.*

50. John Martis, *Philippe Lacoue-Labarthe: Representation and the Loss of the Subject.*

51. Jean-Luc Nancy, *The Ground of the Image.*

52. *Edith Wyschogrod, Crossover Queries: Dwelling with Negatives, Embodying Philosophy's Others.*

53. Gerald Bruns, *On the Anarchy of Poetry and Philosophy: A Guide for the Unruly.*

54. Brian Treanor, *Aspects of Alterity: Levinas, Marcel, and the Contemporary Debate.*

55. Simon Morgan Wortham, *Counter-Institutions: Jacques Derrida and the Question of the University.*